The International Library of Sociology

THE FRAMEWORK OF HUMAN BEHAVIOUR

Founded by KARL MANNHEIM

The International Library of Sociology

SOCIOLOGY OF BEHAVIOUR AND PSYCHOLOGY
In 18 Volumes

I	The Development of Conscience	*Stephenson*
II	Disaster	*Wolfenstein*
	(The above title is not available through Routledge in North America)	
III	The Framework of Human Behaviour	*Blackburn*
IV	Frustration and Aggression	*Dollard, Miller et al*
	(The above title is not available through Routledge in North America)	
V	Handbook of Social Psychology	*Young*
	(The above title is not available through Routledge in North America)	
VI	Human Behaviour and Social Processes	*Rose*
	(The above title is not available through Routledge in North America)	
VII	The Human Group	*Homans*
	(The above title is not available through Routledge in North America)	
VIII	Learning Through Group Experience	*Ottaway*
IX	Personality and Problems of Adjustment	*Young*
	(The above title is not available through Routledge in North America)	
X	Psychology and the Social Pattern	*Blackburn*
XI	The Sane Society	*Fromm*
XII	Sigmund Freud – An Introduction	*Holitscher*
XIII	Social Learning and Imitation	*Miller and Dollard*
XIV	Society and Nature	*Kelsen*
	(The above title is not available through Routledge in North America)	
XV	Solitude and Privacy	*Halmos*
XVI	The Study of Groups	*Klein*
XVII	Theory of Collective Behaviour	*Smelser*
	(The above title is not available through Routledge in North America)	
XVIII	Towards a Measure of Man	*Halmos*

THE FRAMEWORK OF HUMAN BEHAVIOUR

by

JULIAN BLACKBURN

ROUTLEDGE
Taylor & Francis Group

First published in 1947 by
Routledge, Trench, Trubner and Co., Ltd

Reprinted in 1998, 2000 by
Routledge
2 Park Square, Milton Park, Abingdon, Oxon, OX14 4RN

Transferred to Digital Printing 2007

Routledge is an imprint of the Taylor & Francis Group

British Library Cataloguing in Publication Data
A CIP catalogue record for this book
is available from the British Library

The Framework of Human Behaviour
ISBN 0-415-17782-0
The Sociology of Behaviour and Psychology: 18 Volumes
ISBN 0-415-17834-7
The International Library of Sociology: 274 Volumes
ISBN 0-415-17838-X

CONTENTS

PREFACE

Since 1935 when Ruth Benedict, a pupil of Franz Boas, published her book,[1] the culture-pattern theory has been extended and elaborated by Margaret Mead, Gregory Bateson, Otto Klineberg and others.[2] Put briefly, the theory is that just as different peoples have made different selections from all the multitudinous sounds it is possible to make with the human voice, and have woven their selections together into integrated patterns of language, so also different peoples have made different selections out of all the possible ways of living and behaving and have woven them together into integrated patterns of culture. The particular selections which any one group makes gives to that group its characteristic pattern. One group has woven its fabric of culture round the idea of caste, another round the idea of aggression, another round the idea of death, another round money values, and so on. Cultural patterns differ not only in their dominating values, but also in the relative importance which they attach to different subsidiary motives and ways of life, and in the methods by which these subsidiary motives are woven together.

The theory itself may thus be conveniently regarded as an extension into the anthropological field of the approach which Wertheimer, Koffka and Köhler and other "Gestalt" psychologists have developed in general psychology since 1912, though the culture-pattern theory has characteristics and implications all its own.

One of the implications is that when once a pattern of culture has been established, this pattern will in most cases play a more important part in moulding the way in which the members of the community of which the pattern is characteristic will behave, than will the innate differences which the members of that community will possess. It is held that this will be true of the vast majority of the members of that community, although there will always be some whose innate characteristics diverge so far from those which are regarded as of value in that culture-pattern that they will become the deviants and maladjusted people in that community.

A second implication is the notion of the relativity of human

[1] BENEDICT, R., *Patterns of Culture.* London: Routledge, 1935, pp. 291.

[2] MEAD, M., *Sex and Temperament in Three Primitive Societies.* London: Routledge, 1935, pp. 335.

BATESON, G., *Naven.* Cambridge: University Press, 1936, pp. 286.

KLINEBERG, O., *Race Differences.* New York: Harper, 1935, pp. 366.

values. What appears to be quite obvious to one group of people as the most desirable qualities for an individual to possess may seem utterly unimportant or irrelevant to another.

These and other implications of the theory have had an important influence on the ideas of contemporary American anthropologists, but with the notable exception of Professor Pear[1] there have been few psychologists in this country who have examined either the limitations of the theory or its implications from the point of view of the social psychologist.

In my book *Psychology and the Social Pattern* I tried to indicate some of the effects of the theory on the teaching of general psychology. In the present book and in one which will subsequently follow I am concerned to investigate some more of its implications. How far can the theory stand up to the knowledge we have of the mechanism of heredity, and of the part that hereditary characteristics play in human behaviour? How far must it be modified by what we know of sex differences, racial differences, national differences and class differences? These are the topics I am concerned with in this book, though I have felt it to be better to investigate them on their own merits rather than to attempt to link them up to the culture-pattern theory at every succeeding step.

In a subsequent book I shall discuss the development of human beings in society from their birth to maturity and subsequent decline, and the influence on them of their family life, education, occupation, and of the institutions and associations of society. Once again I shall have in mind the aim of determining how far the culture-pattern theory needs to be modified, what are its strengths and what are its limitations in the light of our knowledge of the facts of development in different societies.

* * *

I should like to thank my friends Morris Ginsberg, T. H. Pear, C. H. Waddington and Barbara Wootton for the trouble they have taken in reading over the script of this book and for making a number of useful criticisms and suggestions. Barbara Wootton has also been responsible in one way and another (though occasionally, perhaps, unintentionally) for setting me off in search of a large part of the material.

HADSTOCK, ESSEX *November*, 1945

[1] PEAR, T. H., "The Social Status of the Psychologist and its effect upon his work", *Sociol. Rev.*, 1942, 34, 68–81.
PEAR, T. H., "Psychological Implications of the Culture-Pattern Theory", *Bull. of the John Rylands Library*, 1945, Vol. 29.

CHAPTER I

INTRODUCTION

"Most people in this world seem to live 'in character'; they have a beginning, a middle and an end, and the three are congruous one with another and true to the rules of their type. You can speak of them as being of this sort of people or that. They are, as theatrical people say, no more (and no less) than 'character actors'. They have a class, they have a place, they know what is becoming in them and what is due to them, and their proper size of tombstone tells at last how properly they have played their part."

Tono-Bungay, by H. G. Wells.

The fact that most people betray at any rate a moderate degree of consistency in their behaviour (even though it may be a consistent inconsistency) has enabled authors and novelists to portray individuals who behave in ways which we can all accept as in keeping with their characters. It enables us to recognise as false any behaviour which is out of keeping with the picture of the person which the novelist has drawn. It enables us to guess, quite often with surprising accuracy, how our intimate friends and even how our acquaintances will react to a particular train of events. And it enables skilled interviewers sometimes to select after only a short interview the most suitable person for a particular job (although the ability to do this successfully is not as widespread as many interviewers themselves are inclined to believe). There is no doubt that we expect a certain measure of consistency in character and behaviour from other people, and we are often surprised to encounter elements which are out of harmony with our expectations. Such incongruities have, in fact, been made the subject for many humorous anecdotes.

The outstanding characteristic of man, as Allport[1] emphasizes, is his individuality. He behaves in his own distinctive fashion. "In daily life, in our direct contacts with our fellows, the pre-eminence of individuality is recognised readily enough. During our waking hours and in our dreams people appear to us as definite and individual." On what does this individuality depend? At one extreme there are those who say that it may be explained almost entirely by man's hereditary equipment. Among the

[1] Allport (1) p. 3.

supporters of this view one might place W. C. Dampier Whetham,[1] R. B. Cattell[2] and A. E. Wiggam.[3] Dampier Whetham's book is rather out of date nowadays, and Cattell is primarily concerned with the inheritance of intelligence, but Wiggam surveys a wider field. "The second warning of biology to statesmanship," he writes,[4] "is brief and simple: that heredity and not environment is the chief maker of men; that it is essentially the man, who in the long run makes his environment, much more than it is the environment which makes the man; that man is not a pawn on the chess-board of environment, the football of circumstance and the puppet of chance and change; that he is not a will-o'-the-wisp of fortune, a marionette whose wires are pulled by the hidden hand of doom; that he is not, as the glib reformer has taught you to believe, the helpless victim of the passing education, philosophy and theories of pedagogy of his time; but that, in the germ-cell, from which every man is born, *there are resident* those powerful personal forces by which he can rise in wellnigh any environment and, within the limits of human freedom, exclaim: 'I am the master of my fate; I am the *captain of my soul*' ".

"The social and political import of this warning," Wiggam continues,[5] "is that nearly all the happiness and nearly all the misery of the world are due, *not to environment*, but to heredity; that the differences among men are, in the main, due to differences in the germ cells from which they are born; that social classes, therefore, which you seek to abolish by law, are ordained by nature; that it is, in the large statistical run of things, not the slums which make slum people, but slum people who make the slums; that primarily it is not the Church which makes people good, but good people who make the church; that godly people are largely born not made; that if you want church members you will have to give nature a chance to produce them; that if you want artists, poets, philosophers, skilled workmen and great statesmen you will also have to give nature a chance to breed them."

At the other extreme there are those who hold that the forces of environment far surpass anything which the hereditary equipment of mankind can do. The view is implicit in Pavlov's[6] statement that, "It is obvious that the different kinds of habits

[1] Dampier Whetham (4). [2] Cattell (3). [3] Wiggam (7).
[4] Wiggam (7) p. 35. [5] Wiggam (7) pp. 35–6.
[6] Pavlov (5) p. 395.

based on training, education and discipline of any sort are nothing but a long chain of conditioned reflexes. We all know how associations, once established and acquired between definite stimuli and our responses, are persistently and, so to speak, automatically reproduced, sometimes even although we fight against them."

The doctrine is further elaborated by Watson. "In the case of man," he writes,[1] "all healthy individuals . . . start out *equal.* Quite similar words appear in our far-famed Declaration of Independence. The signers of that document were nearer right than one might expect, considering their dense ignorance of psychology. They would have been strictly accurate had the clause '*at birth*' been inserted after the word equal. It is what happens to individuals after birth that makes one a hewer of wood and a drawer of water, another a diplomat, a thief, a successful business man or a far-famed scientist. What our advocates of freedom in 1776 took no account of is the fact that the Deity himself could not equalize 40-year-old individuals who have had such different environmental trainings as the American people have had."

In a further elaboration of this point of view Watson continues,[2] "Personality changes most rapidly in youth when habit patterns are forming, maturing and changing. Between 15 and 18 a female changes from a child to a woman. At 15 she is but the playmate of boys and girls of her own age. At 18 she becomes a sex object to every man. After 30 personality changes very slowly owing to the fact . . . that by that time most individuals, unless constantly stimulated by a new environment, are pretty well settled into a humdrum way of living. Habit patterns become set. If you have an adequate picture of the average individual at 30 you will have it with few changes for the rest of that individual's life—as most lives are lived. A quacking, gossiping, neighbour-spying, disaster-enjoying woman of 30 will be, unless a miracle happens, the same at 40 and still the same at 60."

It might perhaps be worth pointing out at this point that if the view of Behaviourism about the development of personality is correct, no such miracle could possibly happen, for in order to change the personality, entirely new habits would have to be built up by a long process of training. No sudden changes in personality are therefore possible. Yet one does not have to look far before one encounters numerous cases of sudden "con-

[1] Watson (6) p. 270. [2] Watson (6) p. 278.

versions", or numerous cases of schizophrenic disorders occurring suddenly after a severe emotional shock in people who have shown no previous schizoid tendencies, to realise that Watson's thesis as it stands is untenable.

But perhaps the most famous presentation of Watson's view is contained in this paragraph:[1] "I should like to go one step further now and say, 'Give me a dozen healthy infants, well-formed, and my own specified world to bring them up in and I'll guarantee to take any one at random and train him to become any type of specialist I might select—doctor, lawyer, artist, merchant-chief and, yes, even beggar-man and thief, regardless of his talents, penchants, tendencies, abilities, vocations, and race of his ancestors'. I am going beyond my facts and I admit it, but so have the advocates of the contrary and they have been doing it for many thousands of years. Please note that when this experiment is made I am to be allowed to specify the way the children are to be brought up and the type of world they have to live in."

Watson, it is true, admits that there are certain genetic differences between human beings, that skin colour, eye colour, finger prints, bony structure and so on are hereditarily determined, yet he believes that it is wrong to confuse these *structural* differences with *functional* differences. In fact, however, it is impossible to make this clear-cut distinction between structure and function. Watson himself says,[2] "Infants crawl differently, cry differently, differ in the frequency with which defaecation and urination occur, differ in early vocal efforts, in requirements for food, in the speed and rapidity with which they use their hands—even identical twins show these differences—because they differ structurally and differ slightly in their chemical make-up." Clearly, then, with these structural differences different infants will respond to training in different ways: some will be able to be trained into more satisfactory doctors, lawyers, artists and so on than others. As Carr Saunders,[3] commenting on Watson's claim to be able to train any one to become any type of specialist, says, "But is not this to say, what we all know to our cost to be true, that the most unsuitable people can be turned into doctors, dentists, teachers and engineers?"

Between these two extremes one finds Allport's masterly exposition.[4] No feature of personality is devoid of hereditary

[1] Watson (6) p. 104.
[2] Watson (6) p. 101.
[3] Carr Saunders (2) p. 15.
[4] Allport (1).

influences, but correspondingly no feature of personality is devoid of environmental influences. The same physical environment for individuals with different heredity, as studies of children brought up from birth in institutions shows, does not prevent them from developing widely different personalities, and the same thing is true of widely different physical environments for individuals with the same heredity, as the studies of identical twins reared apart has shown.[1] The personality is a function of heredity *multiplied by* environment. It is not a mere additive relationship: if either heredity or environment were absent there could be no personality.

Nevertheless it may in some cases be possible to suggest which of the two factors is the more important. Differences between siblings brought up in very much the same home environment are probably more largely due to heredity than to environment. Again, dull people are probably more determined by their heredity than are bright people, for the dull people do not have the advantage which their brighter fellows enjoy of being able to learn readily by experience. They therefore do not acquire the capacity of the bright people to adjust easily to different environments, to modify their reactions according to their changing milieux, and so of developing that richness of reaction and sensitiveness to altering stimuli which the brighter people have. They plod on methodically and unimaginatively in the same rut whatever their environment may be.

At birth the individual lacks personality. He is at the one stage in the personality equation where environment = 0 (if one may neglect pre-natal influences on the grounds that they have been acquired in an environment which is different from that in which the individual is going to live). But thereafter an individual's particular genetic constitution meets a particular environment, or a changing environment, or a series of different environments, and the interplay between the constitution and the environment produces his personality, and from an early age the personality that emerges shows a consistency in its development.

All this and much more is elaborated at length and with a wealth of illustration in Allport's book. But to follow Allport along that route would take us far beyond the scope of this book. What we are principally concerned with here is the framework around which human behaviour is built up. This framework is different for every one of us: it depends on our heredity, our sex,

[1] See Chapter III.

our race and nation, and on our social class. This is the framework with which we have to start: this is the framework on which the subsequent influences of family life, education, occupation and all the other institutions, organisations and associations of society will play. Some of us may have a poor heredity compensated by good luck in sex, race, nation and class; others may be lucky, or unlucky, in all these respects. But whatever the combination we may possess, it is here that individual development in society begins, and it is to these topics that we must now turn in order to try to expose what it is we know about them, and in what ways they are likely to affect our subsequent behaviour.

REFERENCES

1. ALLPORT, G. W., *Personality: a Psychological Interpretation.* New York: Holt, 1937, pp. 588.
2. CARR SAUNDERS, A. M., *The Biological Basis of Human Nature.* Oxford: University Press, 1942, pp. 24..
3. CATTELL, R. B., *The Fight for our National Intelligence.* London: King, 1937, pp. 186.
4. DAMPIER WHETHAM, W. C., AND C. D., *The Family and the Nation.* London: Longmans, 1909, pp. 233.
5. PAVLOV, I. P., *Conditioned Reflexes.* Oxford: University Press, 1927, pp. 427.
6. WATSON, J. B., *Behaviorism.* London: Kegan Paul, 2nd Edition, 1931, pp. 308.
7. WIGGAM, A. E., *The New Decalogue of Science.* London: Dent, 1923, pp. 287.

CHAPTER II

THE MECHANISM OF HEREDITY

"He looks a thourough ancester said Ethel kindly.

"Well he was said Bernard in a proud tone he was really the Sinister son of Queen Victoria.

"Not really cried Ethel in excited tones but what does that mean.

"Well I dont quite know said Bernard Clark it puzzles me very much but ancesters do turn quear at times.

"Peraps it means god son said Mr. Salteena in an inteligent voice."—*The Young Visiters*, by Daisy Ashford.

Animals, as we all know, are made of cells, the number of which increases as the animals grow up. In the simpler forms of life all the cells behave more or less alike, but as the process of evolution goes on, and organisms become more complicated, we find that a greater and greater degree of specialization occurs among the cells. Some become modified so as to assimilate food, others so as to assimilate oxygen, others so as to become carriers, others to form muscle and bone and nerve fibres, and so on. This process of differentiation has one major disadvantage that whereas among more primitive forms of life each cell can carry on its existence independently of any other, yet when the high degree of specialisation that we find in the higher forms of life has been attained, each cell, or rather each cell colony, becomes more and more dependent on others for its existence, and if one group dies or ceases to perform its proper functions, all the other cell colonies in the same body suffer.

A cell is a lump of jelly-like substance known as *protoplasm*. In the middle of the cell there is a darker mass of substance called the *nucleus*, the remainder of the cell going by the name of *cytoplasm*. And when cells divide to form new cells, as they continually do—a process known as *mitosis*—the nucleus, if it is examined under the microscope, is seen to form itself first of all into a number of long rod-like bodies known as *chromosomes*. The exact number of chromosomes differs in different species, but it is the same in every member of the same species.

As this process of division occurs, the chromosomes twist and turn about one another, and then at the moment of division each one of them splits in half lengthwise (each chromosome forming two *chromatids*), and exactly one half goes into one of the new cells that is formed, while the corresponding half goes into the other.

One exception to this rule is found, however. When the *germ* cells of animals or plants are formed, the chromosomes instead of splitting up lengthwise form up in pairs. Then as the germ cells are formed one chromosome from each pair goes to each of the new germ cells, so that each germ cell has exactly *half* the number of chromosomes found in the body cells. This process is known as *meiosis.*[1] When two germ cells come together at fertilisation, the two half chromosome sets also come together and make a normal whole set, so that the new individual starts off with exactly the same number of chromosomes as its parents, having received half from its father and half from its mother.

This halving and coming together again of the chromosomes would, therefore, be a perfect mechanism through which hereditary factors might be transmitted to offspring. But the number of chromosomes in any creature is small—in only a few does it exceed a hundred—while the number of different hereditary qualities is very large indeed. Within the chromosomes, therefore, if this is the mechanism of heredity, there might be a number of much smaller units strung out down their length like a string of beads, and each of these units might be principally concerned with one particular characteristic.

Now, although these entities are far too small to be visible even under the most powerful microscope, yet a series of experiments beginning with those of Mendel have led biologists to the conclusion that something of this sort does, in fact, occur. In the first place, Mendel discovered that the characteristics of organisms *are* controlled by definite units which he called *factors* and which have subsequently been called *genes*; and that these units remain entirely unaltered and entirely pure even though they may be considerably mixed with other factors by hybridisation. He also found that some of these factors are *dominant*—that is to say, when they are present they control the particular character they are concerned with—and that others are *recessive*—that is to say, they remain dormant when in the presence of the dominant, although they do not mix or fuse with the dominant but retain their purity.[2] In later generations the recessives may

[1] The whole process of meiosis is more complicated than is represented here but the net result is the same. See Waddington (11) pp. 35–7; Jennings (5) p. 57.

[2] But see below, p. 16. See also Huxley (4) p. 19.

do their work again when once they are freed from the overwhelming influence of the dominant factor, but during their period of submission they remain entirely unaffected.

Working with peas Mendel found that, for instance, yellow seed-colour is dominant over green seed-colour, so that when he crossed a yellow-seeded pea with a green-seeded one he obtained nothing but yellow seeds. When he self-fertilised these hybrid yellow peas, however, he found that the green seeds reappeared. From this he deduced the fact that when the germ cells were formed in the yellow-green hybrids half of them carried the yellow factor and half of them the green. Thus, just as when one tosses two coins a large number of times one finds that two heads together or two tails together appear only half as frequently as one head and one tail, so in the random fertilisation of yellow and green germ cells it is found that one quarter of the offspring will carry nothing but yellow factors, one quarter nothing but green, and the remaining half will carry the yellow and green factors together. Thus, on the average, out of every four peas in the second generation one will be yellow, two yellow-green and one green. But as yellow is dominant over green, the two yellow-green peas will not be distinguishable in appearance from the pure yellow one, and thus we find three apparently yellow peas to each green one, that is to say Mendel's famous 3 : 1 ratio.

On further breeding the pure yellows will breed true and the pure greens will breed true, but the yellow-greens will again give the ratio of three yellows to one green, and so they will go on. See Table 1.

P_1 (parental) generation

Germ cells

F_1 (first filial) generation

Germ cells

F_2 (second filial) generation

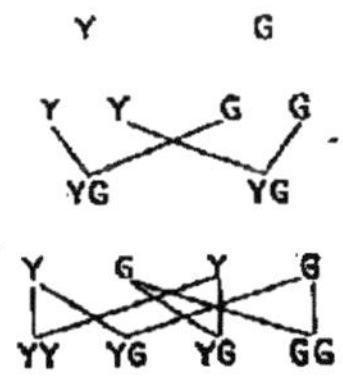

= 3 Yellow, 1 Green.

TABLE 1

With yellow and green seed-colour there is a similarity between the genetic constitution and the appearance of the organism, but in other cases the genetic constitution and the appearance differ. If, for example, a black Andalusian fowl is crossed with a white all the offspring are grey (they are called "blue" by bird fanciers). If these are mated with one another only 50 per cent. of the offspring are grey, the rest being black and white in equal proportions. This indicates that although the appearance (or *phenotypic* character) of the first filial generation was a blending of the appearance of its parents, yet the factors received from the parents (the *genotypic* character) did not themselves blend, for a proportion of them are recovered in the second filial generation, and we can prove by inbreeding that they are uncontaminated, the blacks and the whites continuing to breed true whilst the greys continue to produce offspring in the same proportions as in the F_2 generation.

Similarly, too, among human beings it can be shown that if a negro or negress mates with a white woman or white man, although their offspring, as one would expect on a theory of the blending of elements, are lighter than the former and darker than the latter, yet if these offspring mate, then, in the second generation, instead of—as one would expect on a theory of blending—finding approximately the same colour as their parents in all the offspring, one finds in fact—if the offspring are sufficiently numerous—all degrees of colour from white to black. Again we see that the factors themselves are not affected by the hybridisation.[1] See Figs. 1 and 2.

Mendel next continued his experiments with his "law of independent assortment". This law has had to be considerably revised as a result of more recent experiments. It stated that if two sets of factors segregated in a 3 : 1 ratio, they would, when taken together, segregate in a combination of two such ratios, i.e. in a 9 : 3 : 3 : 1 ratio. For instance, if tallness and shortness and yellow and green colour are taken together, such a proportion will be found in the F_2 generation as is shown in Table 2.

This, however, will only be true if the factors can assort themselves independently. Mendel never happened to come across a case where they did not. The law has, however, now had to be modified because if two factors occur *on the same chromosome* they will not be able to split up in this random way. The whole of a chromosome, it will be remembered, passes into the same germ cell. Two factors on the same chromosome, therefore, cannot pass into different germ cells, and the proportions based on the idea of a purely random mating must inevitably break down. The law of independent assortment,

[1] For a very interesting study of the effects of race mixture in South Africa see Lotsy (6).

FIG. 1. Inheritance of pigment in a cross between negro-white half-castes as conceived in terms of the old Dilution Theory which people believed in before Mendel's work was recognized. From Dahlberg (1), pp. 28–29.

Reproduced by kind permission of Messrs. George Allen & Unwin Ltd.

FIG. 2. Inheritance of pigment in a cross between negro-white half-castes as we now understand the facts. From Dahlberg (1), pp. 28–29.

Reproduced by kind permission of Messrs. George Allen & Unwin Ltd.

[*Face page* 10

therefore, will only be valid in those cases where the factors are situated on different chromosomes. If they are situated on the same chromosome (e.g. in the fruit fly—*drosophila melanogaster*—red or brown eyes and normal or vestigial wings) then they will be linked together during the process of fertilisation, and only two classes in the 3 : 1 ratio will appear in the F_2 generation.

P_1 generation	(T (Tall) (Y (Yellow)	(S (Short) (G (Green)
germ cells	(T (T (Y (Y	(S (S (G (G
F generation	(TS (YG	(TS (YG
germ cells	(T (T (S (S (Y (G (Y (G	(T (T (S (S (Y (G (Y (G

		TY	TG	SY	SG
F_2 generation	TY	Tall Yellow	Tall Yellow	Tall Yellow	Tall Yellow
	TG	Tall Yellow	Tall Green	Tall Yellow	Tall Green
	SY	Tall Yellow	Tall Yellow	Short Yellow	Short Yellow
	SG	Tall Yellow	Tall Green	Short Yellow	Short Green

i.e. 9TY, 3TG, 3SY, 1SG

TABLE 2

This fact has been of enormous importance in enabling biologists to determine which characters are associated with which chromosomes. The tendency of factors to segregate together because they are situated on the same chromosome instead of assorting independently, as on the Mendelian theory, is known as *linkage.*

Linkage is not, however, by any means always complete. In the process of meiosis there is often an exchange of material between the pairs of chromosomes as they are about to divide. This is known as *crossing over.*[1] Part of one chromosome may break off and join up with the other chromosome of the pair in exchange for the corresponding portion of the other chromosome. Thus if a pair of chromosomes before meiosis be indicated by the letters

A A′
B B′
C C′
D D′
E E′
F F′
G G′

[1] Crossing over is now held by Darlington and others to be an invariable and necessary accompaniment of meiosis. See Huxley (4) pp. 133–4.

then when meiosis is complete, if crossing over has occurred, the two chromosomes might be represented by

A′	A
B′	B
C	C′
D	D′
E	E′
F	F′
G	G′

Factors are clearly more likely to cross over together when they are close together on the same chromosome than when they are far apart: it is more likely that the two factors *A* and *B* will cross over together than, say, *A* and *C*, and the determination of cross-over values, i.e. the proportion of times two factors have crossed over together to the number of times they have crossed over separately, has been of the utmost importance in determining the exact position of the different factors on their chromosomes.

These facts, however, were only discovered after Hunt Morgan in America had undertaken his famous experiments on the fruit fly. Morgan chose the fruit fly because it breeds very rapidly, a new generation being produced every fortnight; it is small, so that whole families can be brought up in a bottle; and it also has the very important characteristic of possessing only four pairs of chromosomes, so that it is relatively easy to determine which chromosome is concerned in any changes which occur from generation to generation.

Now an examination of the chromosomes of this fly showed that as the chromosomes twisted and turned about each other in the process of pairing up before the germ cells were formed, they sometimes became very elongated, and that *very occasionally* in these circumstances a piece of the elongated chromosome might break off and become lost, or break off and join up *in a different position* on the other chromosome of the pair. When this happened the offspring showed a characteristic or characteristics which differed from those of its parents. This was all the more noticeable when it happened in one of the chromosome pairs which it was found was not identical—as the other pairs were—in the male and female. For it was discovered that characteristics which were different in the two sexes were associated with this pair, and they could be readily followed up.[1]

Now when this loss of a piece of chromosome or its *translocation* to a different part of the other chromosome of the pair did occur,

[1] See Chapter IV.

the characteristics of the offspring differed according to the *amount* of chromosome that was lost or became translocated. Furthermore, if the same loss or translocation occurred in another individual precisely the same characteristics were affected. In this way it was possible to map out those parts of the chromosome associated with particular characteristics, and the net result of Morgan's experiments was to show that the factors (or *genes*) *are* located side by side in the chromosomes, and that each gene has its own definite position which becomes altered only by accident.

Morgan and his collaborators mapped out the chromosomes of the fruit fly and inserted the relative positions of the genes within them. This showed, as Jennings[1] points out, the following interesting results:—

"1. All the large chromosomes (Nos. I, II, and III) affect all parts of the body: many different functions, health, length of life and the like.

"The small pair (IV), though affecting few characters, still affects diverse parts (eyes, wings, bristles).

"2. Thus particular chromosomes are not limited in their action to any particular part of the body or to any particular function. Each chromosome affects many parts and functions.

"3. It appears clear that each chromosome enters into the process of development, affecting many processes, and doubtless influencing the entire organism.

"4. It is not clear from our present knowledge that the genes are arranged in the chromosomes in any systematic way, with relation to their functions, or to the parts of the body that they chiefly influence. A gene principally affecting eye colour is close to one principally affecting body structure or body colour or structure of the wings, and so on. Genes affecting eye colour occur in many different parts of the genetic system. . . . In a similar way, genes that affect the structure of the wing or of the eye, or that affect the colour of the body, are scattered throughout the four chromosomes.

"Thus if there is any functional system in the arrangement of the genes in the chromosomes, it has not yet been discovered; in the present state of our knowledge it appears as if there were none."[2]

Other workers then tried to map out the genes of other species. Many more difficulties were often encountered in such

[1] Jennings (5) pp. 172–3. [2] But see Huxley (4) p. 86.

work, however, owing to the larger number of chromosomes and greater breeding difficulties.

Apart from translocation there is another way by which changes in characteristics may occur. This is by an alteration in the gene itself. Although genes often go on for many generations without a change, they occasionally do change or *mutate*, as it is called. The mutation rate differs in different genes, in different species and probably in different environmental circumstances, so no reliable estimate of its frequency can be given[1] The cause of mutation has not been discovered. Thus in the fruit fly a race with red eyes may suddenly and unexpectedly produce an individual with white eyes. The gene producing red eye-colour has changed, and some of the descendants of the white-eyed fly may be found to possess the same characteristic. Breeding from them it is possible to produce a race of white-eyed flies which breed true.

Mutation does not alter the position of the gene on the chromosome. The gene for red eye-colour and that for white eye-colour occupy identical positions on their respective chromosomes. Red eye-colour is dominant over white, and when a white-eyed individual is mated with a red-eyed, the offspring, possessing the gene for red eye-colour and the gene for white eye-colour on the appropriate pair of chromosomes, will all have red eyes. An individual of this type, possessing the gene for a particular characteristic in two different forms on the appropriate pair of chromosomes is said to be *heterozygous* for that characteristic. When the genes are of the same form, e.g. in the parents who possess two genes for white eye-colour, or two genes for red eye-colour, the individual is said to be *homozygous* for the characteristic.

A confirmation of the work on mutation and translocation has come from a series of experiments in which both mutations and translocations have been produced artificially. Muller[2] reported that by subjecting fruit flies to X-rays he had been able to induce artificially the same gene mutations and translocations that had arisen "naturally" from time to time in colonies of fruit fly cultures, and that under X-ray bombardment such mutations and translocations and losses occurred 150 times more frequently than they did in flies not subjected to the treatment. Muller's work has subsequently been confirmed by a number of other investigators some of whom have used different forms of short

[1] See Huxley (4) pp. 54 and 358. [2] Muller (9).

wave radiation. Muller further showed that having bombarded parents with X-rays and produced offspring which differed in certain respects from their parents, then, if the chromosomes of these offspring were examined, it could be seen that they showed exactly the same differences from those of their parents as did the chromosomes of those flies which had developed these characteristics naturally.

So far the problem has been discussed as if every different characteristic were controlled by a single gene. Although this is true in some cases (e.g. tallness or shortness, yellow and green colour in peas) yet in the vast majority of cases the characteristics are controlled by a number of genes working together—the *gene complex* or the *genotypic milieu.*[1] In such cases it is found that a change in any one of the genes in the complex upsets the balance of the whole complex and all the genes in that complex are affected.[2]

An example of this is the fur colour of rodents, known as "agouti grey".[3] This is a complicated mixture of colour shades, and on each hair there is a more or less regular pattern of black base and apex with a yellow band between. Agouti animals also have a grey-white belly and undertail and brown eyes. Now this condition has been shown to be due to the presence of at least four pairs of genes which may be called *A*, *B*, *C*, and *D*, all of which have to be present in dominant form for the normal condition to occur. If the pair *A* should become recessive, then an entirely black rabbit with black-brown eyes appears. If *A* remains dominant but the pair *B* becomes recessive, then we get a rabbit which is golden yellow with yellow-brown eyes. If both pairs *A* and *B* change together it is a tortoiseshell with yellow body, dark nose, feet, belly and undertail, and light brown eyes. The gene pairs *C* and *D* are still more interesting. If the pair *D* becomes recessive the rabbit is white with blue eyes—in spite of the fact that it carries both *A* and *B* in dominant form and that therefore (as these are concerned with dark colours) one might think it ought to show *some* dark colour at least. The change in the pair *C* is still more drastic, for in this case the rabbit is a pure

[1] See especially Huxley (4) pp. 19–21, 64–65.

[2] In addition to the fact that any one characteristic may be controlled by several genes there is the fact that any one gene may produce effects on a number of different characteristics (see Huxley (4) pp. 62–3). A modification of such a gene produces changes in the characteristics of different parts of the body. Often, too, when a gene mutates from the normal to a new (usually recessive) form there is a general weakening effect on the whole constitution of the organism.

[3] See Hurst (3) pp. 44–6.

albino with a white coat and pink eyes. Thus the two pairs of genes *C* and *D*, although they produce no colour themselves, are necessary for the proper working of the definite colour genes *A* and *B*, and these can only be effective when *C* and *D* are present in dominant form as well.

Two other facts about the genes which are of special importance to remember when the principles of heredity are applied to human beings are *penetrance* and *expressivity*. In what was said at the beginning of this character it was implied that dominance and recessiveness were absolute things and qualitatively different from each other. It will be remembered that it was said on p. 8 that when a dominant gene is present it controls the particular character it is concerned with, and that a recessive gene remains dormant in the presence of the dominant, but that it does its work again when freed from subjugation to the dominant. It was implied that an individual possessing one dominant gene of a pair must show the characteristic with which that gene is concerned, and similarly that if he possesses a pair of recessive genes he will show the characteristic they are concerned with. This was the original theory, and genes of this type are those which are most usually worked with in the laboratories because they give clear-cut results. Later experiments, however, have shown that dominance and recessiveness are not absolute things.[1] We have already seen that in the fur colour of rodents the dominant genes *A* and *B* can only show their effect when *C* and *D* are present in dominant form too. In this case, as in many others, the dominance is partial and depends on the genotypic milieu. It has further been found that dominance may be influenced by environmental conditions. *Penetrance* therefore is the frequency, measured as a percentage, with which a gene shows any effect at all. Waddington[2] suggests that genes with low penetrance only act for a short period and do not determine the whole course of the development of the particular characteristic they are concerned with. It is, he suggests, as if the genotypic milieu provides a landscape along the main valley of which the character-producing reaction moves. Different genotypic milieux provide different landscapes and thus result in the development of different characters. But a gene acting at the right moment may push the reaction out of the main valley into a side valley with the result that it develops differently. A gene with high penetrance is always or nearly always successful

[1] See also Huxley (4) 68–87. [2] Waddington (11) p. 190.

in doing this: a gene with low penetrance is only occasionally successful.

Furthermore, the characteristics which finally emerge depend not only on the genotypic milieu but also on the influence of environmental conditions. It was mentioned on p. 8, it will be remembered, that in the process of cell division exactly one-half of each chromosome goes into each new cell that is formed. In view of this it may be asked how it is that cells and cell groups can become specialised into different functions, some being modified so as to assimilate food, others to form muscle and bone and nerve fibres and so on. The answer to this problem is not very clear at present, though a number of interesting facts which are known may be relevant. One is that there are regional differences in the properties of the cytoplasm so that when, after cell division, the different nuclei have become enclosed in cytoplasm having different properties the reaction of the same heredity in different environments produces a different final result. Furthermore, although the chromosomes divide exactly in half, the cytoplasm does not. Once again the same heredity in a different environment may produce a different end result. In the early cell divisions, however, this differentiation of function is not fixed. If the two cells produced after the first division of the fertilised egg are separated, they do not produce two half individuals but two whole individuals. The same thing is true of any one of the cells produced after the next few divisions. Thus the development of each cell depends on its environment, that is to say on its relation to other cells. After a while, however, though sometimes not till quite a late stage of development, things become more fixed and any given cell, though it contains all the necessary genes, no longer produces an entire individual nor even a given part of the organism. It cannot function without its environment of other cells.[1]

One of the most startling effects of environmental conditions on development is seen in the axolotl. This is "a large salamander that has large red external gills, a tail flattened sidewise, for swimming, and other features that fit it for living in the water. It lives in the water thus all its life, becomes mature, produces eggs and young; there it dies; it is never anything but an axolotl.

"But if it is fed on thyroid, it undergoes a transformation comparable to that which changes a tadpole into a frog. It loses its gills; its body form becomes greatly changed, its tail is

[1] See Waddington (10) and (11) pp. 192–206 and Jennings (4) pp. 181–9.

no longer flattened sidewise. It becomes fitted for living on land, crawls out on the land, and is the creature known as Amblystoma. It now lives all the rest of its life under this guise, as a land animal; it becomes mature in this condition, producing eggs and young.

"Of greatest interest from our present point of view is the fact that external conditions may induce this same transformation. If the axolotl is driven to come out on land under certain conditions of temperature and the like, it transforms into an Amblystoma, just as it does when fed on thyroid. There is little doubt that what these conditions do is to cause the thyroid of the animal to discharge its secretion into the blood, and that this induces the transformation. Thus the axolotl may have either one or two very diverse sets of characteristics, depending on the conditions that it meets during development."[1]

Again in *Gammarus chevreuxi*, a small shrimp-like animal, there are two types of eye-colour, black (dominant) and red (recessive). But the individuals with the recessive red eyes may be enabled to adopt the dominant form if they are kept at high temperatures. Finally, in the fruit fly, there is a mutation which produces a giant form. But the percentage of giants which will emerge from the larvæ (though they all possess the requisite genetic constitution) will depend on the amount of food that is available during a particular period of their development.

In all these cases therefore the following general principles apply:

"1. What are really inherited (passed on from parent to offspring) are the genes; that is, certain materials that in certain combinations and under certain conditions give rise to certain definite characteristics.

"2. With the same original genes, different environmental conditions may induce the production of diverse characteristics.

"3. Also, with the same environmental conditions, different genes may induce the production of different characteristics.

"4. The same difference in characteristics that is in one case produced by diversity of genes is in other cases produced by diversity of environment."[2]

Expressivity may be distinguished from penetrance in that whereas penetrance is a measure of the frequency with which a gene shows any effect at all, expressivity is a measure of the amount of effect shown by a gene. In the giant form of the fruit fly, the expressivity hardly varies at all. Any giants that do

[1] Jennings (4) pp. 227–9. [2] Jennings (4) p. 237.

emerge are all of much the same size in relation to the normal. In other cases expressivity and penetrance may both vary, sometimes in the same direction, sometimes in opposite directions. Expressivity, like penetrance, may depend on the genotypic milieu or on the effect of the environment. In the latter case it is of great importance in the light it may throw on the problem of the relative importance of heredity and environment. But for reasons that will be mentioned in a moment it has unfortunately so far been found to be more useful in studying the fruit fly than in studying human beings.

REFERENCES

For a more detailed account of the principles and of the mechanism of heredity the reader is referred to the following works which have already been mentioned in the text.

1. DAHLBERG, G., *Race, Reason and Rubbish.* London: Allen & Unwin, 1942, pp. 240.
2. FORD, E. B., *The Study of Heredity.* Home University Library. London: Butterworth, 1938, pp. 255.
3. HURST, C. C., *Heredity and the Ascent of Man.* Cambridge: University Press, 1935, pp. 138.
4. HUXLEY, J., *Evolution: the Modern Synthesis.* London: Allen & Unwin, 1942, pp. 645.
5. JENNINGS, H. S., *Genetics.* London: Faber, 1935, pp. 351.
6. LOTSY, J. P., "Hybridisation among human races in South Africa", *Genetica,* 1928, *10,* pp. 315.
7. MAGUINESS, O. D., *Environment and Heredity.* London: Nelson, 1940, pp. 216.
8. MORGAN, T. H., *The theory of the gene.* Oxford: University Press, new edition, 1928, pp. 376.
9. MULLER, H. J., "Artificial transmutation of the gene", *Sci.,* 1927, *66* pp. 84-7.
10. WADDINGTON, C. H., *How Animals Develop.* London: Allen & Unwin, 1935, pp. 127.
11. Waddington, C. H., *An Introduction to Modern Genetics.* London: Allen & Unwin, 1939, pp. 441.

CHAPTER III

HEREDITY AND HUMAN PROBLEMS

"That is Mary Ann Fudge my grandmother I think said Bernard she was very well known in her day.

"Why asked Ethel who was rarther curious by nature.

"Well I dont quite know said Bernard but she was and he moved away to the next picture."—*The Young Visiters*, by Daisy Ashford.

Whereas the fruit fly breeds rapidly, has large families, and is capable of being brought up in large numbers in small bottles, human beings breed slowly, have very small families, and though they can be brought up on bottles they cannot be brought up in them. Whereas, too, the fruit fly has only four pairs of chromosomes, man has 24 pairs. Again, whereas with the fruit fly it is possible to bring up large numbers with the same or with different genetic constitutions in the same or in different environments, it is only on very rare occasions—i.e. identical twins—that human beings have the same genetic constitution, and the environments in which they are brought up can certainly only very rarely, and perhaps never, be made identical. It therefore follows that the genetic study of human beings presents an extremely difficult task. What certain knowledge there is derives from medical practice. On the psychological side (apart from certain mental abnormalities which are due to organic causes) very little *certain* knowledge exists about the mechanism of heredity.

It is usual to classify inherited characteristics into three types: there are those called *hereditary* which may be traced from generation to generation: there are those that are *familial*, occurring in several members of the same family and then disappearing: and there are those that are sex-linked, appearing only in members of one sex—in human beings the male—but being transmitted by members of the other. Sex-linked characteristics will be discussed in Chapter IV. Here we shall be concerned with hereditary and familial characteristics.

Physical Characteristics[1]

"Hereditary" traits are transmitted by a dominant gene, and in the most usual case of a person showing the characteristic

[1] See Gates (12) and Baur, Fischer and Lenz (1) pp. 213–404.

mating with someone who does not, it will be expected on Mendelian laws that half the offspring will show the characteristic. This follows from the assumption that a person possessing the characteristic is of the heterozygous type DR (where one of the genes of the appropriate pair is dominant and the other is recessive) while the normal mate is homozygous RR (both genes being recessive). Therefore half the offspring will be heterozygous and the remainder homozygous.

Diabetes insipidus, a condition in which a large amount of urine of low specific gravity is excreted, *brachydactyly* in which the middle phalanx of the fingers is missing, and split foot are among the diseases which are transmitted in this way. In many other cases, however, the difference between the expected number of individuals showing the characteristic and the number actually reported as showing it is significantly greater than would be

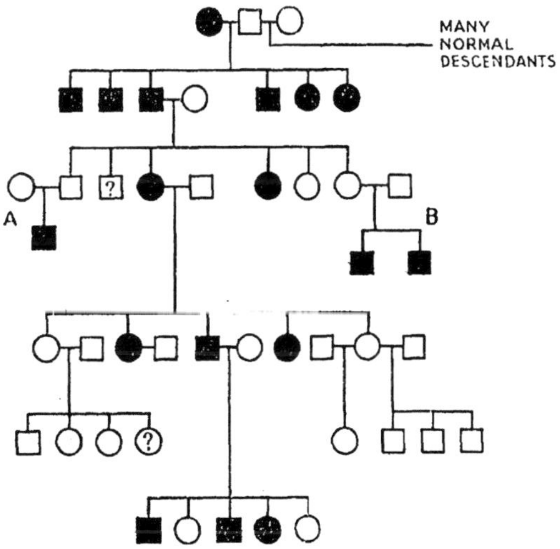

FIGURE 3. A Pedigree of Glaucoma.
Males = Squares. Females = Circles.
Affected individuals Black.
(From Waddington (33) p. 329.

expected by chance. There are a number of reasons that may account for this. Methods of diagnosis may overlook some actual cases: information about ancestors must usually be obtained at second hand: paternity is not always beyond dispute: the penetrance of the gene may not be perfect: and a particularly favourable environment may delay or prevent the development of the characteristic. For instance, the above pedigree of *glaucoma* (a condition characterised by an ever-increasing degree of intra-ocular tension due partly to congestion which if untreated leads to blindness) illustrates that at the two points *A* and *B* it is transmitted by non-affected individuals. It is probable, therefore, that its penetrance is not complete, as otherwise the individuals at *A* and *B* would have shown it.

Characteristics determined by recessive genes cannot be followed from generation to generation. They may appear suddenly among the brothers and sisters (siblings) of the same family, and then, unless there is cross-cousin marriage, they usually disappear—perhaps for many generations. They will occur if the two parents both possess the gene in the heterozygous form. In these circumstances neither parent shows the characteristic, but one-quarter of their offspring (as with the F_2 generation of tall and short peas) will show it. The person possessing the characteristic will be of the type RR. His parents will both be DR, and if he mates with a homozygous type DD, which he will be most likely to do if the characteristic is a rare one, none of his offspring will show the characteristic which may remain submerged for generations.[1]

The statistics of familial incidence are especially subject to error, since, as human families are small, those with only one affected are not uncommon, and these are often omitted since they do not show any "familial" incidence. A mathematical correction can, however, be made for this.

One of the best tests for a recessive gene arises in cross-cousin marriage, for there is then a much better chance that the two parents will be heterozygous for the gene in question, and that therefore affected offspring will appear. Consequently, if the parents of those showing the trait show an excess of inbreeding over that which is characteristic of the population as a whole, that is very good evidence that the trait in question is caused by a recessive gene.

[1] The fact that recessive characters may remain submerged for many generations and then, after two parents bearing the recessive character have mated, reappear in their original form is one of the strongest pieces of evidence *against* the theory of the inheritance of acquired characters. When it reappears the recessive character is the same as it was when it first appeared, in spite of the fact that generations of organisms may have exhibited a different bodily character. The theory of the inheritance of acquired characters assumes that modifications in bodily characteristics may affect the germ plasm, with the result that such modifications are transmitted to offspring. The germ cells therefore, according to this theory, are derived from the body cells instead of, as most biologists believe, the body cells being derived from the germ plasm, a small portion of which persists unchanged in the germ cells. For a discussion of the evidence on this theory see, e.g. Holmes (17) Chapter 2, Maguinness (23) Chapter 9, Huxley (18) pp. 457 *et seq.*

Albinism is one of the traits in which this has been shown most clearly. In over 600 pedigrees of albinism which were collected by Karl Pearson and his associates cross-cousin marriage occurred in 17 per cent. of the cases (whereas among the community as a whole it is less than one per cent.).

One such pedigree is shown in Fig. 4. In this there is an unusual amount of inbreeding—between an uncle and his niece

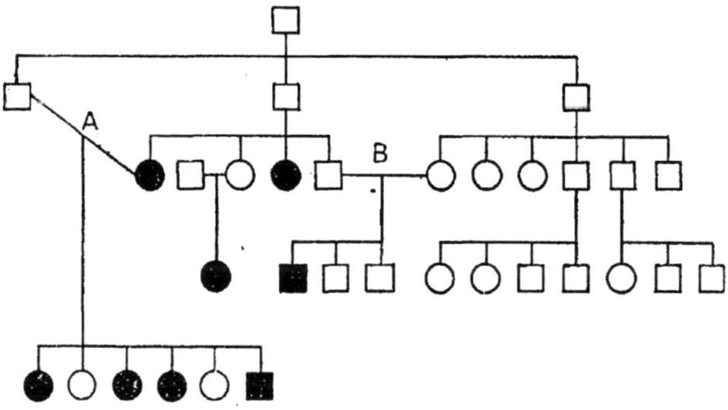

FIGURE 4. A Pedigree of Albinism.
Males = Squares. Females = Circles.
Affected individuals Black.
(From Waddington (33) p. 330.)

at *A*, and between first cousins at *B*—which accounts for the appearance of the characteristic in successive generations.

Among many of the other physical diseases which are hereditarily determined there are grounds for believing that endemic cretinism and diabetes mellitus have a hereditary background, though their exact mode of inheritance is still a matter of dispute.[1] The unpleasant characteristics of both these diseases can, however, be removed by taking an appropriate drug—thyroid extract for cretinism and insulin for diabetes. There are cases on record of repulsive looking imbeciles being transformed into attractive and intelligent children by the administration of thyroid—showing in fact nearly as much difference in their appearance and behaviour as axolotl and amblystoma[2]—though in other cases the extent of the improvement is not so phenomenal. So far as diabetes is concerned there is no need to emphasize the importance of the discovery of insulin. It has transformed a formerly rapidly fatal disease into little more than a major

[1] See Blacker (3) Chapters 14 and 15, or Gates (12) p. 222 and p. 309.
[2] See pp. 17-18 above.

inconvenience. The lives of diabetics can be prolonged to the normal span.

In both diabetes and cretinism the appropriate drug must be taken at intervals during the whole of the rest of the patient's life if the ill effects of the disease are to be avoided. Nevertheless in both cases a genetic weakness has been compensated by means of an external or environmental influence. It is not perhaps too fantastic a supposition that as our knowledge of the nature of genes and their inter-relationship expands more means may be discovered either to control them or to provide substitutes such as are already provided for the cretin or the diabetic, so as to counteract the deleterious effects of other genetic weaknesses. We could then call in environmental agencies to make good these other hereditary deficiencies. But the rich train of exciting possibilities which this line of thought discloses must be left to the biologist to establish and to the novelist to embroider.

Mental Characteristics

When we turn to the inheritance of mental characteristics we at once pass on to far more difficult and debatable ground. Part of the difficulty arises, as Lewis[1] points out, because of difficulties in classification. "Elsewhere in medicine," he writes, "one can distinguish between health and disease readily enough in practice. But in psychiatry it is often a dubious decision. According to the canon of general medicine the delimitation of a disease depends on the discovery of common etiology, pathology and clinical features. In psychiatry this is hard to apply. It is not necessary here to discuss in detail the different conceptions of disease, types of reaction, symptom-complexes, etc., nor the grounds of classification, but it must be pointed out that identical "physical causes" may lead to very different clinical manifestations of mental disorder, and that the main psychoses are akin to, and at times indistinguishable from, variations or types of personality that fall within the bounds of statistical normality."

In the second place the role played by the environment is often of much greater importance than it is in purely physical characteristics. To quote Lewis[2] again: "It is not to be supposed that in all the manifestations of mental disorder heredity is prepotent. Environmental influences also account for much. A hereditary predisposition (genotype) may be present, but it

[1] Lewis (22) p. 87. [2] Lewis (22) pp. 88–9.

may never manifest itself, i.e. never have opportunity to appear as a phenotype (e.g. a person with tendency to senile dementia may die young; a psychopath, or potential addict, may not have access to drugs, etc.) or it may manifest itself in a mitigated form (e.g. cyclothymia, quarrelsome disposition) or it may appear in an apparently different form (*petit mal*, *grand mal*, epileptic absences). These variations may be conditioned by the more or less accidental happenings of individual life. It is often peculiarly difficult in mental disorder (especially the non-bacterial, non-traumatic varieties) to assess the part played by environmental influences: the schools differ, and there are few rules that all accept. But the geneticist in this field is hamstrung if he does not know the environment may enter into the clinical picture; he must be able to reckon with it. It is not sufficient to be aware that 'G.P.I. is caused by the spirochaete', or that 'traumatic psychosis is secondary to gross injury of the brain'. Between such conditions as Huntington's chorea or amaurotic idiocy, which are almost entirely dependent on hereditary factors, and, at the other end of the scale, hysteria and reactive psychoses, there is a great and subtle range of varying interplay between the transmitted and the experienced. By clinical analysis, a great deal of light may be thrown upon these relations."

In the third place the hereditary basis of mental characters is highly complicated and almost certainly polygenic, and this may account for the inability to express mental inheritance along such regular Mendelian lines as can be employed when inheritance depends on a single dominant or a pair of recessive genes. As Huxley[1] emphasizes: "Up till very recently the chief data in human genetics have been pedigrees of abnormalities or diseases collected by medical men. And in collecting these data, medical men have usually been obsessed with the implications of the ideas of 'character-inheritance'. When the character has not appeared in orthodox and classical Mendelian fashion they have tended to dismiss it with some such phrase as 'inheritance irregular', whereas further analysis might have shown a perfectly normal *inheritance* of the gene concerned, but an irregular *expression* of the character, dependent on the other genes with which it was associated and upon differences in environment."

With these difficulties and limitations in mind, let us turn first of all to the two main types of functional mental disorder, namely manic-depression and schizophrenia. Of these Slater[2]

[1] Huxley (18) p. 20. [2] Slater (29) p. 555.

suggests that cyclical manic-depression is inherited by a single dominant gene. The evidence in favour of this hypothesis is by no means undisputed, and even if it is correct the dominance must be of low penetrance for only 20 to 30 per cent. of the carriers of the manic-depressive gene ever develop the psychosis. That there is some hereditary basis for it, however, appears to be clear from Table 3[1] which shows the liability of various relatives of manic-depressives to be themselves manic-depressive:

	Per cent.
Parents	11.5
Brothers and sisters	9.1
Children	9.5
Nephews and nieces	2.3
Uncles and aunts	5.0
First cousins	2.5

TABLE 3. Liability of various relatives of manic-depressives to be manic-depressive themselves. (From Slater (29) p. 556.

As the average individual risk is about 0.4 per cent. it will be seen that even the more remote relatives of manic-depressives are more liable to the illness than the average.

So far as schizophrenia is concerned Slater suggested in 1938[2] that it depended in the first place on a single recessive gene. Yet here again the environment, or low penetrance, or the genotypic milieu prevent the appearance of the disorder in some of those who possess the homozygotic RR constitution. It is estimated that only about two-thirds of those who possess the double recessive gene ever develop the disorder, though many or the rest may show abnormalities of temperament, such as being reserved, fanatic, eccentric, suspicious, sensitive, frigid, etc. It is believed that some of those who possess the heterozygotic constitution may also show these temperamental difficulties. If this is so it indicates that so far as these temperamental abnormalities are dependent on hereditary factors the recessiveness of the gene is not absolute.

Of the uniovular twins of parents both of whom are schizophrenic four out of five become schizophrenic, instead of all as would be expected. Here the difference between the observed and the expected results can only be due to the environment. The difference between two-thirds and four-fifths, therefore,

[1] Slater (29) p. 556. [2] Slater (29) p. 556.

provides some measure of the effect of the genotypic milieu acting alone.

Although Slater suggested in 1938, after carefully weighing the evidence available at that time, that schizophrenia depended in the first place on a single recessive gene, he now[1] in the light of further evidence inclines to a theory of dominance. It may be, therefore, that in view of the apparent irregularity of the inheritance, some theory based on a polygenic origin will in the future elucidate the mode of inheritance of schizophrenia and also of manic-depression.

The inheritance of Huntington's chorea is fairly well established as being due to a single dominant gene. It is a disorder in which there is progressive mental deterioration accompanied by the exaggerated restlessness and nervous twitchings characteristic of the choreatic patient. Out of 744 cases collected from the literature by Julia Bell[2] only six were found in which there did not seem to be inheritance from one or other parent. Even in this case, however, the penetrance of the gene is less than 100 per cent. for instead of finding 50 per cent. of the parents, siblings and children of the Huntington's patient themselves exhibiting the disorder, only about 35 per cent. of them do.

But most of the characteristics we are concerned with in human beings are characteristics which vary continuously and do not have clear-cut distinguishing features such as those just mentioned.[3] Most of them, like height, weight, intelligence and many personality and temperamental characteristics, show all degrees from the lowest to the highest, and arbitrary lines have to be drawn between those people whom we choose to regard as falling into one class and those who fall into another. This introduces a further complication into the investigation of the hereditary mechanism on which they may be based, in addition to the three which were mentioned on pp. 24-5. In such cases other methods than those we have been considering up to now are more usually employed for determining their hereditary nature. These methods depend on the calculation of the degree of association which exists between the possession of varying degrees of the characteristic in question and varying environmental circumstances, or between the possession of varying degrees of the characteristic in question and varying degrees of

[1] Slater (30) pp. 26–8. [2] Bell (2) quoted by Slater (29) p. 554.
[3] For a discussion of the genetic relationship between continuous and discontinuous variability in characters see Huxley (18) pp. 64–8.

heredity—from identical twins who possess the same hereditary endowment, through parents and children, brothers and sisters, uncles and aunts, nephews and nieces, first cousins, and other more remote relatives. The correlation coefficient[1] is the measure most often employed.

The Inheritance of Intelligence

Let us take as an example of this approach some of the work that has been done on the inheritance of intelligence. This started with the work of Galton[2] before the statistical device of the correlation coefficient had been developed. Galton observed that certain families appeared to monopolise most of the existing ability. His standard of eminence was the position achieved by not more than one person in 4,000, and he took the most eminent man in each family in this category and then tried to find out how many of such a person's near and more remote relatives had also attained the degree of eminence he demanded. He considered eight occupational classes—judges, statesmen, commanders, literary men, scientists, poets, artists and clergymen—and worked out the percentage of eminent men of near or remote relationship to the most eminent man in the family. He worked it out for each of the classes separately and for all of them combined. The latter gave the result seen in Table 4.

These percentages are far greater than one would expect by chance, and they also indicate that the more distant the relative the smaller the percentage of eminence becomes. There is a fair indication, therefore, that hereditary factors are playing an important part in determining eminence.

[1] A correlation coefficient is a statistical measure of the amount of concomitant variation between two variables. When two variables exhibit perfect direct concomitant variation, when, that is to say, an increase in one is accompaned by a corresponding increase in the other, or a decrease in one by a decrease in the other, the correlation coefficient between them is expressed by the figure +1. Such, in physics, is the relationship between the pressure and the temperature of a gas when the volume is held constant. When two variables exhibit perfect inverse concomitant variation, when, that is to say, an increase in one is accompanied by a corresponding decrease in the other, or a decrease in one by an increase in the other, the correlation coefficient between them is expressed by the figure −1. Such, in physics, is the relationship between the pressure and the volume of a gas when the temperature is held constant. When two variables exhibit no concomitant variation, when, that is to say, an increase in one is sometimes accompanied by an increase and sometimes by a decrease in the other in no regular or recognizable way, the correlation coefficient between them is 0. An imperfect degree of direct or inverse concomitant variation between two variables is expressed by a correlation coefficient between 0 and +1, or between 0 and −1 respectively, the coefficient approximating more closely to +1 or to −1 as the degree of concomitant variation approaches perfection.

[2] Galton (10)

	Per cent. of eminent male relatives of eminent men
Father	31
Brother	41
Son	48
Grandfather	17
Uncle	18
Nephew	22
Grandson	14
Great-grandfather	3
Great-uncle	5
First cousin	13
Great-nephew	10
Great-grandson	3
All other more remote relatives	31

TABLE 4. (From Galton (10) p. 308.)

The pedigree of the Darwin family gives even more convincing evidence, for Charles Darwin married his first cousin, and this inbred family was extremely successful. This can be seen in Fig. 5.

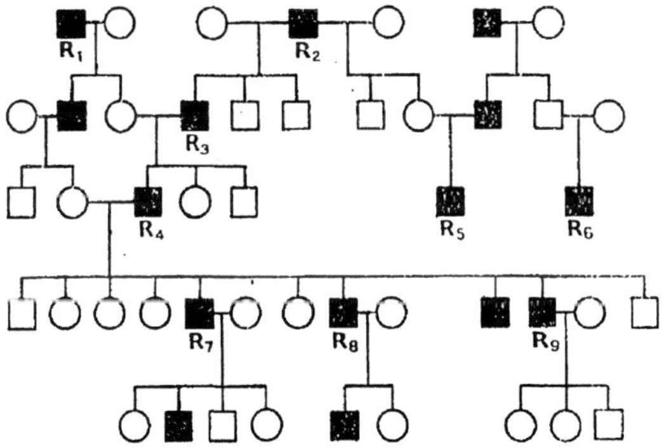

FIGURE 5. A Pedigree of the Darwin Family.
Males = Squares. Females = Circles. R = Fellows of the Royal Society. Black Squares = Men distinguished for intellectual eminence. The intellectual eminence of the women has not been assessed.
R_1 = Josiah Wedgwood, founder of Wedgwood Pottery Works.
R_2 = Erasmus Darwin.
R_4 = Charles Darwin.
R_5 = Francis Galton.
R_7 = George Darwin, astronomer and mathematician.
R_8 = Francis Darwin, Botanist.
(From Holmes (17) p. 153.)

This gives a strong presumption in favour of the view that

hereditary factors play some part in determining the tendency of eminence to run in families. There is, however, in these investigations, no way of telling how far the favourable environment, opportunities and influence in which the children of the eminent people were brought up influenced their own rise to eminence, or how many of them would have failed to appear in the statistics of eminence had they been born with the same heredity but in circumstances as unfavourable as they were in fact favourable. The fact that in Galton's investigation the percentage for fathers is significantly smaller than that for sons may be some indication of the effect of these factors.

The same thing is true of the investigations which have been made into degenerate families. The Jukes[1] family, the descendants of five sisters or half sisters, have been traced through five generations and their cost to the State in pauperism, disease and crime has been estimated at 1½ million dollars within 75 years. Another family, the Kallikak family,[2] consists of two branches, one through the legitimate marriage of the original Kallikak with a girl of normal intelligence, and the other through his illegitimate union with a feeble-minded girl. The descendants of the former have had normal or even distinguished careers, those of the latter have been a source of constant charge on the State through feeble-mindedness. Even with the Kallikaks, however, the influence of environment cannot be entirely ruled out. It is at all events possible that if a similar environment had been provided for the offspring of the feeble-minded girl and of the girl of normal intelligence, or, alternatively, if the original Kallikak had married the feeble-minded girl and had had illicit relations with the normal girl, different results might have been obtained.

The more recent investigations have made use of intelligence tests and correlation coefficients. One of these is Terman's investigation[3] into geniuses which has been proceeding for twenty years and is still continuing. A representative sample of children aged between 2 and 13, most of them being 10 at the time the investigation began, who possessed an Intelligence Quotient[4] of more than 140, were intensively studied and have

[1] See Dugdale (8) and Estabrook (9).
[2] Goddard (13). [3] Terman (31).
[4] The Intelligence Quotient ("I.Q.") is a statistical measure of an individual's level of intelligence. Normal intelligence is represented by an I.Q. of 100, different degrees of mental defect by an I.Q. of 70 or less, and genius by an I.Q. of 140 or more, though authorities differ in the precise classifications they adopt. See my *Psychology and the Social Pattern*, chapter 5.

been followed up two and seven years after the original investigation was made. Terman, like Galton, found that a large proportion of the children had eminent relatives and that they came from family stocks with decidedly superior intellectual endowment. He believes that the intellectual eminence of the children he studied must be due very largely to heredity, for the educational opportunities in California are as nearly as possible equal, and yet his gifted group of children came predominantly from parents in the higher occupational levels. Previous studies had only considered the parents themselves: here, however, the same thing is found to be true of children while they are still subject to the very similar influence of the school environment. In some cases, too, evidence (apart from intelligence tests) for the existence of superiority appeared in the pre-school period in a wealth of miscellaneous information, desire to learn to read, and intellectual curiosity far beyond that of most children.

Terman used the Stanford-Binet Intelligence test and a verbal group intelligence test to measure the children's level of intelligence. Such tests—in fact all tests, though some more than others—are to some extent influenced by the environment in which the child is brought up.[1] This is of particular importance when the group contains, as did this group, children of different racial origins as well as of different social origins. But environmental influence can certainly not account for the whole of Terman's results, though the exact genetic basis for the inheritance of intelligence is still obscure.[2]

A number of investigations have indicated that the average I.Q. of rural children is lower than that of urban children. In one of these Cattell[3] measured representative children aged 10 in an industrial city (Leicester) and a rural area (Devon). He found that the average intelligence of the rural group was seven points lower than that of the urban. He also found that the average size of family in the rural area was greater than that in the urban, and that in both urban and rural groups the number of children per family was progressively greater as one passed down the intelligence scale from the top to the bottom. Cattell calculates that in both urban and rural groups the I.Q. is declining at a rate of about three points per generation, and that if the

[1] See Chapter V. See also my *Psychology and the Social Pattern*, Chapter 5.

[2] Recently Fraser Roberts (26) has produced evidence to indicate that the inheritance of intelligence rests on a multi-factorial basis.

[3] Cattell (5).

present differential birth rates continue then in 300 years half the population will be mentally defective.

Cattell used a non-verbal test to measure the I.Q.s and such tests are much less subject to environmental influence than are verbal tests, though even with non-verbal tests we do not know at present the exact extent to which urban and rural environments, home influences and schooling may affect the scores. It is possible, however, that it might be sufficient to account for a difference of only seven points in I.Q. between the urban and rural groups. In the first place identical twins reared apart, as we shall see below, have shown as the result of environmental agencies a far bigger difference than 7 points in their I.Q.s. Secondly Gordon[1] has found that the average intelligence of canal boat children is normal up to the age of six years. It is only after six that apparent backwardness appears—a backwardness which increases with age until the average I.Q. at the age of nine is only 70. Gordon suggests that a possible explanation is that the canal boat children, by missing schooling, miss the opportunity to acquire the abilities that would enable them to answer the intelligence test questions satisfactorily. Similarly, therefore, a possible reason for the apparent backwardness of rural children when compared with urban children may be that the design of the tests makes it easier for urban than for rural children to do them. Most intelligence tests do not present the two groups with situations of equal difficulty. Tests are usually overweighted with items that are more familiar to city children than to rural children. In their early years more of the play activities of city children compared with country children give the city children an advantage in the tests which are used. This is indicated in the work of Jones, Conrad and Blanchard[2] who found that urban and rural groups differed very widely in their answers to tests of the Stanford-Binet scale, and that many of the age assignments of the tests were too high when the tests were applied to the rural group. The tests which showed the greatest degree of rural inferiority were those which involved the use of paper and pencil, those which involved familiarity with coins, trams and other things more common in the city, and those which involved the use of abstract terms. They conclude that the rural inferiority in I.Q. is due, partly at any rate, to environmental factors and to the form of standardisation of the scale, and that a rural child on moving into a town would increase his intelligence test scores

[1] Gordon (14). [2] Jones, Conrad and Blanchard (19).

merely as a result of changed environmental conditions. In addition to this, the less regular school attendance in rural areas, together with the level of education and teaching which is superior in the towns, would further increase the advantage of the town children.

In order to get a better comparison between the intelligence of urban and of rural children, therefore, it would be necessary, as Schwesinger[1] suggests, to design two tests, one a principally rural test, the other a principally urban test, to standardise them on their respective groups, and then to compare the performance of the urban children on the rural test with that of the rural children on the urban test. No one has yet done this, although Shimberg[2] has shown that if two types of test are prepared, one dealing with items more familiar to urban children and the other with items more familiar to country children, although the country children will score lower than the urban children on the first test, the urban children will score lower than the country children on the second. Either group might therefore be ranked as superior according to the type of test employed.

Studies of the Effect of Environmental Influences

(a) *Foster Homes*

The problem how far environmental agencies affect the result of investigations into the inheritance of intelligence is one that is constantly recurring. Different approaches have been made in order to try to measure their influence. One approach is by way of foster homes into which children have been adopted at an early age. The intelligence of the adopted children is compared with that of the foster parents' own children, and if the correlation coefficient between them is significantly higher than zero—for there is normally, of course, no correlation between the intelligence of unrelated individuals brought up in heterogeneous environments—it may indicate that the influence of the foster home has been to make the intelligence of the adopted children approximate to some extent to that of the own children. Alternatively, a coefficient of correlation between the intelligence of the adopted children and of the foster parents may be calculated, and again if it is higher than zero similar deuctions may be made.

Unfortunately there is a serious difficulty. Unless it can be

[1] Schwesinger (27). [2] Shimberg (28).

shown with absolute certainty that there is no selective placement, that is to say no attempt to match a particular child with a particular type of foster home, the foster home studies are valueless. The point has been investigated. On the surface it appears that there is no selective placement so far as intelligence is concerned. The foster parents make their choice (in so far as they make a choice at all) on the apparently more important grounds of sex, age, personal appearance, health and winning ways. But sometimes a process of unconscious matching apparently does occur. A number of children had their intelligence tested before being adopted into foster homes. Subsequently the intelligence test scores were correlated with a rating of the approximate mental level of the foster homes, and there was found to be a coefficient of correlation of $+0.34$ between them, a coefficient which is approximately the same as that between mid-parent (i.e. average of father and mother) and own child, indicating a fair degree of unconscious matching. It is at present a matter of dispute whether the selective placement is sufficient to account for all the intellectual resemblance between foster parents and their adopted children, or whether some part of the resemblance may be attributed to environmental influences within the foster home. Further experiments are necessary before an adequate answer to the question can be given.

(b) *Orphanage studies*

The intelligence of unrelated children brought up in the same environment of an orphanage or institution provides the material for a second group of investigations. The implicit assumption is that institutional life provides an *equivalent* environment. In point of fact it is unlikely to do so, for some children will be temperamentally better fitted than others to endure a life devoid of parental care and attention. Some may find their emotional outlets and expression are adequately satisfied, but others may feel the need of a personal affection so deeply that they rebel against the institution's authority. There is no justification for assuming that the environmental forces are equivalent in the two cases. The difficulty is borne in mind by Lawrence,[1] but she attempts to dispose of it by maintaining that in a sufficiently large institution any one type of environmental impact is unlikely to be confined to children having a similar heredity, and that

[1] Lawrence (21).

consequently the unevennesses in the environment will tend to cancel out.

One of the first questions to be answered was the relationship between the intelligence of the children and the occupation of their parents. Other investigations (e.g. the American Army data of Yoakum and Yerkes[1] and Cattell's investigation[2]) have shown that, although there is some overlap, the I.Q. of people in more skilled occupations is higher than that of those in less skilled. Furthermore, Duff and Thomson[3] who gave an intelligence test to 13,419 school children showed that there was a tendency for the level of intelligence of the children to vary according to their father's occupation, the average I.Q. of the children of professional men being 112.2, that of the children of engineers 102.9, that of the children of labourers 96, and so on. Lawrence obtained similar results with a control group of London elementary school children. However, such findings might be attributable to differences in the home environments: what happened in the case of the children who were removed from their parents' care when they were less than twelve months old and who were reared in a uniform institutional environment? Figures were calculated for the correlation between, on the one hand, the intelligence of boys and of girls and, on the other, the father's occupation, the mother's occupation and the mid-parent occupation (obtained by averaging the occupations of father and mother). For purposes of calculation the occupations were divided into five levels, ranging from the professional class to dock labourers and paupers. In every case some positive correlation was found, indicating that "though a child has never lived with its parents, it is likely, other things being equal, to have a slightly higher intelligence if it comes from one of the so-called upper classes than if it is the child of labouring people".[4] At the same time the difference between the average levels of intelligence of the five occupational classes was not as great in the institutional cases as it was in the control group of elementary school children, a fact which indicates that environmental factors have some influence on the test results.

The second question to which Lawrence addressed herself was the effect of a longer period at home before admittance to an institution on the relationship between intelligence and parental occupation. One of the institutions she studied admitted children

[1] Yoakum and Yerkes (34).
[2] Cattell (4).
[3] Duff and Thomson (7).
[4] Lawrence (21) p. 32.

at different ages. She therefore calculated separately the correlation between the children's intelligence and the parents' occupation for the children admitted before they were 3 and for those admitted after they were 3. The groups were small (38 boys and 30 girls in the "under 3" group, 57 boys and 50 girls in the "over 3" group) but the results showed a difference that is probably significant statistically. Those who remained longer in their homes provided a higher correlation than those who entered the institution before they were 3. In addition to this, an increasing correlation between intelligence and class with increase in age was found in the control group of elementary school children who remained at home. Both these findings point to the influence of environment on test results.

On the other hand when Lawrence investigated the effect of improved environmental conditions, by separating from her cases those who had come from extremely bad homes (e.g. N.S.P.C.C. cases, Guardians' cases, and the children of insane, drunken or criminal parents) and divided them into groups according to the proportion of their lives that had been spent in the institution, she found practically no increase in intelligence score for the longer periods of residence.

Finally, she worked out coefficients of variation in order to discover whether a period of continued residence in the uniform institutional environment caused the test results to become less variable. Previous work suggested that it might be so, but no control groups of non-institutional cases had been used so it was not possible to know whether the diminished variability was due to the effect of the institutional environment or to a normal decrease in variability after the age of 10. Lawrence therefore provided a control group from outside of similar ages to the children in the institution, and when she compared the two groups she found little difference in variability between them.

Crissey[1] has obtained results which indicate quite clearly that the average level of intelligence in an institution has a determining effect on the level of intelligence of new admissions to the institution. If the new admissions are above the average level of the institution they show losses in intelligence after periods of residence, whereas if they are below it on admission they remain constant or show slight gains. The group level appears to set the pattern of mental development of its members. Crissey found that in environments where the mental level was that of the

[1] Crissey (6).

normal or slightly subnormal child the level of intelligence of children who were classified as borderline or feeble-minded when they entered the institution tended to show slight gains or to remain constant after a period of residence in the institution, whereas that of children who were classified as normal or superior showed consistent losses in intelligence. The extent of the losses appears to be related to the original degree of divergence from the average of the institution and also to the length of residence in the institution. When children were transferred from institutions designed for normal and dull-normal children to other institutions designed primarily for children at the feeble-minded level, their mental development after transfer tended to follow the pattern set by the level of the new environment. Those classified at the time of their transfer as borderline or feeble-minded showed consistent losses after their transfer, whereas the level of intelligence of high-grade imbeciles remained fairly constant.

When sixteen dull-normal and normal children were transferred from institutions for the feeble-minded to institutions designed for normal and dull-normal children, their level of intelligence again followed the pattern set by the new environment. All the children with I.Q.s of 93 or over showed losses after transfer, while those with I.Q.s of 85 or less showed gains. The gains or losses were more closely related to the mental level of the children than to their age at transfer.

When the influence of corresponding periods of residence in environments of different average mental level on the intelligence of children was investigated, it was found that the children showed losses in intelligence in the schools for the feeble-minded whereas children of the same age and mental level in orphanages retained their original mental level. In the orphanages the children represented the lower mental levels whereas in the schools for the feeble-minded they were among the higher mental levels. Again, therefore, it appears that the differing mental levels of the two environments set different patterns of environmental stimulation.

(c) *The Study of Twins*

If we now turn to the question of twins we find in the first place the penetrating observations and easy-flowing pen of Sir Francis Galton.[1] Galton's work was anecdotal rather than

[1] Galton (11).

experimental. New techniques have been developed since his day: statistics were only in their infancy then, and no intelligence tests had been invented when his observations were recorded.

Galton divided his twins into two classes, those who were exceedingly alike and those who were exceedingly unlike in childhood. He had 35 pairs in the first class and 20 in the second. Subsequent work has confirmed the division of twins into two classes, the monozygotic, developed by the splitting of a single fertilised egg, and the dizygotic, developed by the fertilisation of two separate eggs. The former are known as identical twins; they are always of the same sex and are closely similar in physical characteristics. The latter are known as fraternal twins; they may be of the same sex or of different sexes, and they are no more alike than are ordinary siblings. It is probable, in the light of subsequent knowledge, that most of Galton's like twins were identical and most of his unlike twins were fraternal.

Galton obtained information about the twins by sending out a questionnaire asking for certain particulars. One of the questions asked for anecdotes regarding mistakes made between the twins by their near relatives. "The replies are numerous, but not very varied in character. When the twins are children, they are usually distinguished by ribbons tied round the wrist or neck; nevertheless the one is sometimes fed, physicked, and whipped by mistake for the other, and the description of these little domestic catastrophes was usually given by the mother, in a phraseology that is somewhat touching by reason of its seriousness. I have one case in which a doubt remains whether the children were not changed in their bath, and the presumed A is not really B, and *vice versa*."[1]

"No less than nine anecdotes have reached me of a twin seeing his or her reflection in a looking glass, and addressing it in the belief it was the other twin in person."[2]

Galton describes an almost photographic similarity of the illnesses undergone by the twins. "Both twins were apt to sicken at the same time in no less than nine out of the thirty-five cases. Either their illnesses, to which I refer, were non-contagious, or, if contagious, the twins caught them simultaneously; they did not catch them the one from the other."[3]

Other anecdotes concerned the similarity in the mental processes of the twins. "No less than eleven out of the thirty-five cases testify to this. They make the same remarks on the same

[1] Galton (11) pp. 157–8. [2] Galton (11) p. 158. [3] Galton (11) p. 162.

occasion, begin singing the same song at the same moment, and so on; or one would commence a sentence, and the other would finish it. . . . One of the most curious anecdotes that I have received concerning this similarity of ideas was that one twin, A, who happened to be at a town in Scotland, bought a set of champagne glasses which caught his attention, as a surprise for his brother B; while, at the same time, B, being in England, bought a similar set of precisely the same pattern as a surprise for A."[1]

Galton, having presented his evidence concerning the extreme similarity between the twins in their early youth, next tried to discover what effect the different environments that the twins experienced after they were grown up and had separated, had on their former similarity. He found that in some cases the similarity of body and mind persisted, apparently unaltered, up to old age, in spite of very different conditions of life; and in no case was a subsequent dissimilarity attributable entirely to the combined action of a number of small environmental differences. When dissimilarity in later life was mentioned, its cause was most commonly attributed to some illness such as scarlet fever or typhus that had struck one of the twins but not the other. Galton concludes from this part of his investigation that nature is much stronger than nurture.

He then turns to the converse side of the picture, the case of the unlike twins, in order to discover whether a similarity of environment causes any decrease in dissimilarity between them. He quotes remarks about thirteen of the twenty cases he collected, and says that the evidence in each case is absolutely concordant. For instance: "(1) One parent says: 'They have had *exactly the same nurture* from their birth up to the present time; they are both perfectly healthy and strong, yet they are otherwise as dissimilar as two boys could be, physically, mentally, and in their emotional nature' . . . .(13) The father remarks:—'They were curiously different in body and mind from their birth'. The surviving twin (a senior wrangler of Cambridge) adds:—'A fact struck all our school contemporaries, that my brother and I were complementary, so to speak, in point of ability and disposition. He was contemplative, poetical, and literary to a remarkable degree, showing great power in that line. I was practical, mathematical, and linguistic. Between us we should have made a very decent sort of a man'."[2] Galton says he had other cases in which the

[1] Galton (11) p. 165. [2] Galton (11) pp. 170–71.

word "complementary" was used, and that he had not a single case in which an original dissimilarity became less through identity of nurture.

His conclusion is, ". . . the impression that all this evidence leaves on the mind is one of some wonder whether nurture can do anything at all, beyond giving instruction and professional training. It emphatically corroborates and goes far beyond the conclusions to which we had already been driven by the cases of similarity. In those, the causes of divergence began to act about the period of adult life, when the characters had become somewhat fixed; but here the causes conducive to assimilation began to act from the earliest moment of the existence of the twins, when the disposition was most pliant, and they were continuous until the period of adult life. There is no escape from the conclusion that nature prevails enormously over nurture when the differences of nurture do not exceed what is commonly to be found among persons of the same rank of society and in the same country. My fear is, that my evidence may seem to prove too much, and be discredited on that account, as it appears contrary to all experience that nurture should go for so little."[1]

Such were Galton's conclusions. He was handicapped by having to rely on anecdotal descriptions of behavioural characteristics, and by having no quantitative measurements to present. It was nearly 20 years before Thorndike[2] presented quantitative results, nearly 40 years before an experiment was done on twins by Merriman in which intelligence tests were used,[3] and 50 years before a really far-reaching investigation was published.[4] No investigation since Galton's day, with the exception of Lange's,[5] has presented such emphatic conclusions in favour of nature and against the part played by nurture. Lange investigated 30 pairs of twins—13 identical and 17 fraternal—one of whom had been imprisoned. He decided to find out what had happened to the other twin of the pair. Among the 13 pairs of identical twins the second member of the pair had also been imprisoned in 10 cases, but in 3 cases had remained clear of the law: but among the 17 pairs of fraternal twins the second had been imprisoned in only 2 cases; in 15 cases this had not occurred. Lange's conclusions were still more strongly emphasized when he investigated the type of crime committed by the different members of each identical twin pair, e.g. in one case both were committed to prison

[1] Galton (11) p. 172. [2] Thorndike (32). [3] Merriman (24).
[4] Newman, Freeman and Holzinger (25). [5] Lange (20).

for financial swindles, in another case they were both committed for petty larcenies, in another for burglary and so on. Lange also stated that in all these cases, in addition to the type of crime being similar, the criminal careers began at about the same time and that the behaviour of the members of the pairs in court and in prison had been closely similar. But in the two cases of fraternal twin pairs both of whom had been imprisoned there was no such parallelism as there was in the identical twin pairs, either in type of crime, age of beginning a criminal life or general behaviour.

A rather different type of conclusion from Lange's emerges from the recent study of twins by Newman, Freeman and Holzinger which illustrates the changed technique and controlled experimental methods which have been developed since Galton's day.

In the first part of their study they collected 50 pairs of identical twins and 50 pairs of fraternal twins, each pair having been brought up together. In the case of fraternal twins only those of the same sex were selected, so as to avoid any possible complication to the test results caused by a difference of sex. The ages of the twins ranged between 8 and 18 years.

The authors point out that one of the difficulties in investigations of this kind consists in the diagnosis of twins as identical. Criteria must be used which are independent of the tests subsequently employed to discover the effects of environmental differences. If, for example, it is required to learn the comparative effect of environment on the mental ability of identical and fraternal twins, the classification of the twins into identical and fraternal must be independent of any tests of mental ability. This would appear to be obvious, but it is rather overlooked by Hirsch[1] who eliminated from his similar group of twins those who were unlike in school work, and from his dissimilar group those who were judged to be similar scholastically and mentally. His subsequent conclusion that heredity is six times as important as environment in determining intelligence quotients is, therefore, somewhat invalidated by his method of exaggerating the difference between his groups.

A provisional diagnosis of identical twins was made by Newman, Freeman and Holzinger on the basis of ten criteria. Only those who satisfied all the requirements of the ten criteria were classified as identicals. Some of these criteria were that the

[1] Hirsch (15).

twins must be so strikingly similar in general appearance that they were likely to be mistaken for each other; that they must have the same eye, hair, and skin colour; the same facial features, and the same types of teeth; that they must possess stronger cross-resemblance than internal resemblance in most of the details of finger and palm patterns (that is to say, one hand of one twin must be more like one hand of the other twin than like his own other hand). The provisional diagnosis on the basis of these criteria was then reviewed in the light of a detailed examination of palm prints and finger prints.

All the twins were then subjected to a number of physical and mental measurements. The physical measurements included height, weight, head length and width, cephalic index (i.e. the ratio between head length and breadth), and total number of finger ridges. The mental measurements included the Binet I.Q., Otis I.Q., Stanford Educational Age, and Woodworth-Mathews neurotic inventory. The differences given by each pair of twins for each of these measurements were noted and were compared with those given by a control group of ordinary siblings of like sex. For instance, the Binet I.Q. mean pair difference for identical twins was 5.9 points, for fraternal twins 9.9. points, and for ordinary siblings 9.8 points. In general it was found that the differences between pairs of fraternal twins were very similar to those of ordinary siblings, but that those between pairs of identical twins were much less, and fell into a different kind of distribution. The contrast was sharper in the physical than in the mental measurements.

The fact that the average difference in I.Q. between identical twins reared together was as high as 5.9 points gives some measure of the effect of environmental agencies on, or the unreliability of, intelligence test results. For identical twins have the same heredity and any differences in their scores must be due either to environmental agencies or to the unreliability of the methods of measurement. With this in mind one can see that Cattell's difference of only 7 points in I.Q. between urban and rural groups[1] is not so startling as it appears to be at first sight.

Next it was decided to compare the intra-pair differences of the younger with those of the older twins. The identical twins were divided into two groups at the age of 13 years 2 months, giving 26 pairs in the younger and 24 in the older group. The fraternal twins were divided at the age of 13 years 4 months,

[1] See pp. 31-2.

giving 21 pairs in the younger and 31 in the older group. The mean differences for height, mental age, and educational age are shown in Table 5.[1]

	IDENTICAL			*FRATERNAL*		
	Younger	*Older*	*Difference*	*Younger*	*Older*	*Difference*
Height	1.7 cm.	1.6	−0.1 ± 0.3	4.4	4.4	0.0 ± 0.7
M.A.	7.2 mo.	9.8	2.6 ± 1.3	10.6	19.9	9.3 ± 2.2
E.A.	5.5 mo.	7.5	2.0 ± 1.1	6.5	15.1	8.6 ± 1.5

TABLE 5. Mean differences of pairs of twins in height, mental age and educational age. (From Newman, Freeman and Holzinger (25) p. 107.

It will be seen that there is no difference in height between younger and older identical twins, and none between younger and older fraternal twins. In the case of mental age and educational age, however, the position is different. There is a slight but probably not significant difference between younger and older identical twins, and a large and significant difference between younger and older fraternal twins. The authors conclude, "Since it is reasonable to assume that the environment of fraternal twins is less alike than that of identical twins, the decrease in likeness in mental traits of the one type coupled with the maintained likeness of the other is in harmony with the view that such traits are affected by the environment."[2]

So far we have been concerned with the differences between identical twins, fraternal twins and siblings, all of whom had been reared together. In the second part of the investigation, however, information was collected about 19 pairs of identical twins who had been reared *apart* since infancy. It is this which puts the whole investigation into a class by itself. In Lange's investigation and in the other investigations on twins each pair of identical and each pair of fraternal twins had been reared together. The assumption underlying Lange's work and these other investigations is that the environment for fraternal twins reared together is as similar as that for identical twins reared together. It was on this that Lange's conclusion about the importance of nature as

[1] Newman, Freeman and Holzinger (25) p. 107.
[2] Newman, Freeman and Holzinger (25) p. 111.

against nurture mainly rested. But the assumption itself is a questionable one, as we have just seen, and the conclusions which Newman, Freeman and Holzinger draw as the result of their investigation of identical twins reared apart put the relative roles of environment and heredity in a rather different perspective.

Seventeen of the nineteen pairs had been separated before they were two years old, one pair when about three years old, and one when about six years old. The last case was included because of the abnormally large difference in environment to which the twins were subsequently subjected. The length of separation ranged between 11 and 53 years.[1] The pairs differed in the degree of completeness of separation: in a number of cases it was complete for a long period, and in six of these the twins learnt of the other's existence only after one of them had been mistaken for the other; in other cases the twins had visited each other or had at least communicated with each other at intervals during the period of separation. The distance of separation, both geographical and social, also varied. At the time of testing the age range of the twins was from 11½ to 59 years, with an average age of 26 years.

The same methods of diagnosing identicals were employed as were used in the first part of the experiment, and they were subjected to similar physical and mental measurements. In addition, ratings were made by five skilled investigators of the educational, social and physical differences to which the members of each pair had been exposed. The agreement between these ratings which were made by the five investigators independently was very close, the correlation between them being greater than +0.9.

Correlation coefficients were then worked out between the intra-pair differences on the various tests and the educational, social and physical differences to which they had been exposed. In the physical measurements the only significant correlation was between weight and differences in physical health (+0.6). There was no significant correlation between any of the other physical measurements (height, head length, head width, and cephalic index) and social, educational or physical differences.

Significant positive correlations were found between intelligence test differences and differences in education. The largest was +0.79 between Binet I.Q. and educational differences;

[1] No correlation was found between the length of separation and the extent of the intra-pair differences in mental, temperamental and physical measurements.

those for the Otis I.Q., the International test, and the American Council test were lower than this, but still significantly positive. Some of these differences were remarkably high, e.g. seven of the nineteen pairs differed by 10 points or more in Binet I.Q. (three points higher than the average difference between urban and rural groups found by Cattell in spite of the identical heredity of the twins); one pair showed a difference of nearly 4 years in mental age, two of 3 years, three of 2 years, and four of between 1 and 2 years.

Significant positive correlations were also found between the intelligence test differences and differences in social environment.

Taken as a whole the correlations indicated that significant differences in intelligence and scholastic achievement, as measured by tests, were produced by differences in educational and social environment. It should be mentioned, however, that most of the relationship was produced by relatively few cases with an extreme difference in their environments. If, for example, four cases with a very big difference are eliminated, the correlations between the Binet I.Q., Otis I.Q., International test and Stanford Achievement test on the one hand and the educational differences on the other are reduced from +0.79, +0.55, +0.46 and +0.91 to +0.41, +0.25, +0.34 and +0.45 respectively. These correlation coefficients are not statistically significant, though they are all consistent and positive. It would therefore appear that the influence of environment is more than proportionately greater as the difference is increased.

A comparison was then made between the twins reared together and the twins reared apart. Comparing the separated identicals with the unseparated identicals significant mean differences were found for weight, intelligence, and scholastic achievement. The separated identicals were, in fact, found to be as different as unseparated *fraternals* in weight, intelligence and educational achievement. Thus, "the environmental factor, operating alone in the case of separated identical twins, can produce differences as great as or greater than those produced when both hereditary and environmental factors operate within twin families, as in the case of fraternal twins."[1] The fact that the average differences in intelligence between pairs of separated identicals were as great as those between pairs of unseparated fraternals disposes of the claim, as Hogben,[2] who has himself made important studies of twins, points out, that I.Q. differences

[1] Newman, Freeman and Holzinger (25) p. 348. [2] Hogben (16) p. 152.

in a mixed population are a satisfactory measure of inborn capability.

In some of the physical characteristics, particularly height and head measurements, the separated identicals were more alike than fraternals: they were approximately as much alike as the unseparated identicals.

Finally a formula was worked out for estimating the part played by environment in the differences between the different groups. Comparing identicals reared together with identicals reared apart the share of environment in determining *weight* was stated to be 87 per cent.: comparing identicals reared together with fraternals it was only 21 per cent. Similarly, for Binet I.Q. the two methods gave 59 per cent. and 31 per cent. respectively; for Otis I.Q. they gave 64 per cent. and 16 per cent.; and for Stanford Achievement 87 per cent. and 36 per cent. respectively. That is to say, "if the environment differs greatly as compared with heredity, the share of environment in determining traits which are susceptible to environmental influence is large. If, on the other hand, there is large genetic difference and small environmental difference, the share of heredity is relatively large. This is what makes the solution of the question as to the relative share of the two sets of factors indeterminate. We would have to specify what degree of genetic difference is to be compared with what degree of environmental difference. We may, however, add this statement that differences in the environment which actually sometimes occur, as exemplified in our separated pairs, are sufficient to produce differences in weight, ability, and behaviour large enough to overshadow the genetic differences which occur between siblings."[1]

Before this chapter is concluded there is one other observation which it is necessary to make. The environmental influences we have been considering are differences in the conditions under which the individuals live. With identical twins, however, there is sometimes another difference which it is difficult to know whether to class as environmental or hereditary. Identical twins have the same genetic constitution: every cell in the body of both members of every identical twin pair has an identical chromosome outfit with the same dominant and recessive genes and the same genotypic milieux, etc. But physical differences between them may occur as the result of the stage of development which has been reached before the single egg divides into two. This often does not occur until many cells are present—sometimes

[1] Newman, Freeman and Holzinger (25) p. 359.

not until the differentiation of bodily parts has already begun. If it occurs early the twins will have a greater degree of physical resemblance than if it occurs later: for if it does not occur until the different parts of the embryo are already beginning to become differentiated, then one of the twins will be right-handed and the other will be left-handed, for the twin derived from the right half of the embryo will have his right side already beginning to develop whereas his left half will have to start developing anew, and the converse will be true of the twin derived from the left half of the embryo—his left side will be already slightly developed while his right side formed at the plane of separation will have to begin developing afresh. Right-handedness in one member of a pair of indentical twins and left-handedness in the other is, in fact, sometimes found. Left-handedness occurs with much greater frequency among twins than in the general population. Similar differences may also be found between the members of twin pairs, e.g. the hair whorl of Marie Dionne is clockwise while that of the other four Dionne children is counter-clockwise.

In addition to, and connected with, this fact there is another which may be of still greater significance in the psychological and temperamental development of identical twins. If the embryo does not divide until a late stage one side (usually the right) will be somewhat ahead of the other in development. After division, therefore, the twin who comes from the right half may be more developed and more vigorous than the one who comes from the left half. This also has been observed in fact. And such a difference in vigour and development may have considerable psychological and temperamental consequences, e.g. the more vigorous may become the leader, the less vigorous the led, etc.

Finally there is a suggestion that in some cases the division of the embryo may not be exactly equal. Here again though each twin possesses the same genetic constitution the one developed from the larger half may be bigger and more vigorous than the other with psychological and temperamental consequences similar to those mentioned in the preceding paragraph.

Taken together these facts may have important results. In spite of their identity in genes the members of identical twin pairs may already be unequal at the moment when they begin their post-natal life, and their consequently different reactions to the environments they meet as their lives continue, even though they are brought up in the same family and in as similar a manner as may be, may increase and exaggerate these early differences.

REFERENCES

1. BAUR, E., FISCHER, E., AND LENZ, F., *Human Heredity.* London: Allen & Unwin, 1931, pp. 734.
2. BELL, J., *Treasury of Human Inheritance.* London: 1934, Vol. 4, part 1.
3. BLACKER, C. P. (*Ed.*), *The Chances of Morbid Inheritance.* London: H. K. Lewis, 1934, pp. 449.
4. CATTELL, R. B., "Occupational Norms of Intelligence," *Brit. J. Psychol.*, 1934, *25*, 1-28.
5. CATTELL, R. B., *The Fight for our National Intelligence.* London: King, 1937, pp. 186.
6. CRISSEY, O. L., "Mental Development as related to Institutional Residence and Educational Achievement". *Univ. of Iowa Studies: Studies in Child Welfare*, 1937, *13*, pp. 81.
7. DUFF, J. F., AND THOMSON, G. H., "Social and geographical distribution of intelligence in Northumberland". *Brit. J. Psychol.*, 1923, *14*, pp. 192-8.
8. DUGDALE, R. L., *The Jukes.* New York: Putnam, 1910, pp. 120.
9. ESTABROOK, A. H., *The Jukes in 1915.* Washington: Carnegie Institute, 1916, pp. 85.
10. GALTON, F., *Hereditary Genius.* London: Macmillan, 1914, 2nd edn., pp. 379.
11. GALTON, F., *Inquiries into Human Faculty and its Development.* Everyman Library, London: Dent, 1928, pp. 261.
12. GATES, R. RUGGLES, *Heredity in Man.* London: Constable, 1929, pp. 385.
13. GODDARD, H. H., *The Kallikak Family.* New York: Macmillan, 1921, pp. 121.
14. GORDON, H., "Mental and Scholastic Tests among Retarded Children". *London: Board of Educ., Educational pamphlets*, No. 44, 1923, pp. 92.
15. HIRSCH, N. D. M., *Twins: Heredity and Environment.* Cambridge, Mass.: Harvard University Press, 1930, pp. 159.
16. HOGBEN, L. T., "Genetic variation and Human Intelligence", *Proc. 7th Internat. Genet. Congress.* Cambridge: University Press, 1939 pp. 147-52.
17. HOLMES, S. J., *Human Genetics and its Social Import.* New York: McGraw Hill, 1936, pp. 414.
18. HUXLEY, J., *Evolution: the Modern Synthesis.* London: Allen & Unwin, 1942, pp. 645.
19. JONES, H. E., CONRAD, H. S., AND BLANCHARD, M. B., "Environmental Handicap in Mental Test Performance", *Univ. Calif. Publ. Psychol.* 1932, *5*, pp. 63-99.
20. LANGE, J., *Crime as Destiny.* London: Allen & Unwin, 1930, pp. 250.
21. LAWRENCE, E. M., "An Investigation into the Relation between Intelligence and Inheritance", *Brit. J. Psychol. Monog. Suppl.*, 1931, *16*, pp. 80.
22. LEWIS, A. J., Chapter 4 in Blacker (3) above.
23. MAGUINESS, O. D., *Environment and Heredity.* London: Nelson, 1940, pp. 216.
24. MERRIMAN, C., "The Intellectual Resemblance of Twins", *Psychol. Rev. Monog. Suppl.*, 1924, *33*, No. 152, pp. 58.
25. NEWMAN, H. H., FREEMAN, F. N., AND HOLZINGER, K. J., *Twins: A Study of Heredity and Environment.* Chicago: Univ. Chicago Press, 1937, pp. 369.
26. ROBERTS, J. A. FRASER, "The Resemblance in Intelligence between Sibs", *Ann. Eugen.*, 1940, *10*, pp. 293-312.
27. SCHWESINGER, G., *Heredity and Environment.* New York: Macmillan, 1933, pp. 484.

28. SHIMBERG, M. E., "An Investigation into the Validity of Norms with special reference to Urban and Rural Groups", *Arch. Psychol.*, 1929 No. 104, pp. 84.
29. SLATER, E., "Mental Diseases: Heredity", in *The British Encyclopædia of Medical Practice*, 1938, Vol. 8, pp. 552-63.
30. SLATER, E., "Genetics in Psychiatry", *J. Ment. Sci.*, 1944, *90*, pp. 17-35.
31. TERMAN, L. M., *Genetic Studies of Genius.* London: Harrap, Vol. I, 1925, pp. 648; Vol. iii, 1930, pp. 508.
32. THORNDIKE, E. L., "Measurement of Twins", *Arch. of Philos., Psychol., and Sci. Meth.*, 1905, No. 1.
33. WADDINGTON, C. H., *An Introduction to Modern Genetics.* London: Allen & Unwin, 1939, pp. 441.
34. YOAKUM, C. S., AND YERKES, R. M., *Intelligence Tests in the American Army.* London: Sidgwick & Jackson, 1920, pp. 303.

CHAPTER IV

SEX DIFFERENCES

"A lady is smarter than a gentleman, maybe.
She can sew a fine seam, she can have a baby,
She can use her intuition instead of her brain,
But she can't fold a paper on a crowded train."

PHYLLIS MCGINLEY.

CHROMOSOMAL BALANCE [1]

It was mentioned on p. 12, Chapter II, that when chromosome pairs are examined it is found that there is one pair which is *not* identical in males and females: all the others are. In mammals the female has two large chromosomes, called "X", while the male has one large X and a small "Y" chromosome. Thus all female mammals have two X chromosomes in every body cell and one X in every egg cell, and in this respect the egg cells are alike. The males of mammals, on the other hand, have an X and a Y chromosome in every body cell, and consequently produce two kinds of sperm, one half of which carry an X chromosome and one half a Y. [2]

The first thing to bear in mind, therefore, is that if the male sperms carrying the X chromosome meet and fuse with the female eggs carrying X chromosomes, the resulting offspring will all be females, since they will carry two X chromosomes. On the other hand if the male sperms carrying Y meet female eggs, all of which carry X, the progeny from these will all be male, since they will bear XY chromosomes. See Table 6.

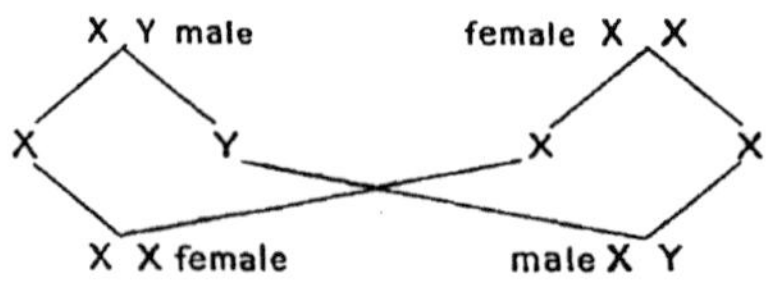

TABLE 6. (From Ford (5) p. 57.)

The inheritance of genes carried by the sex chromosomes is very curious in many cases. In the fruit fly, for instance, it was

[1] See, especially, Ford (5) Chapter 3.

[2] In some organisms, e.g. birds, the males are XX and the females XY.

found that if a female with white eyes was mated with a male with red eyes, instead of the progeny all having dominant red eyes, as would be expected, the sons were all white-eyed while the daughters were all red-eyed. Thus the characters crossed over to the opposite sex. It was then discovered that the red and white eye-colour was carried by a gene in the X chromosome, but that there was no corresponding gene in the smaller Y chromosome. Thus when a white-eyed female is crossed with a red-eyed male the female produces egg cells which *all* bear the white gene in their X chromosomes, while in the male only the X-bearing sperms carry the red gene—the Y chromosome having no corresponding gene. Therefore when the X eggs meet the Y sperms there is no dominant red gene to mask the white and all the sons (XY) have the white eyes of the mother. But when the red-bearing X sperms join the white-bearing X eggs the red will dominate the white and all red-eyed daughters (XX) will result.

In mankind colour-blindness and hæmophilia (bleeding) are the best known sex-linked characteristics. These are transmitted as in the above example. Females do not usually show the characteristic, though they may possess it in a recessive form. That is to say, the gene for the characteristic is dominated by the corresponding gene in the other chromosome of the pair, and the individual does not show the characteristic. The only way in which females could show it would be for *both* the genes concerned in the appropriate pair of chromosomes to be present in recessive form. As a matter of fact no certain case of hæmophilia in women has yet been discovered and there is some evidence[1] that the female sex hormone may counteract its influence even if she were to possess the two recessive genes. Alternatively it may be that the possession of one hæmophilic gene has a general weakening effect on the constitution,[2] and that the possession of two would have a lethal effect so that the female possessing them would never mature. Colour-blind women are, however, found.

If, however, the females possess the characteristic but do not show it, i.e. if they are heterozygous for the characteristic, then half their egg cells will have the X chromosome containing this recessive gene, and if one of these happens to be fertilised by a Y carrying sperm, the characteristic will appear in the resulting male child. If, however, one of the X chromosomes containing the *dominant* is fertilised by a Y-carrying sperm, the resulting male

[1] Waddington (15) p. 332. [2] See Chapter II, p.15, footnote 2.

child will not show the characteristic. That is why not all the sons of a hæmophilia-carrying mother are hæmophilic.

It is important at this point to carry our knowledge of the mechanism of sex determination a stage further. Further work has brought out the fact that although the presence of the Y chromosome may sometimes be necessary for the fertility of males it is not an active sex determinant. In some organisms it is completely absent. Sometimes, too (perhaps through some failure in cell division), there have been found *females* who have a Y chromosome as well as their two X chromosomes; or males with an extra Y. The sex of these creatures is not affected.

Now, observe what happens in successive generations. A male gets his X chromosome from his mother. In him it determines his sex as a male. Yet this very same chromosome was one of those which helped to determine his mother's sex as a female. Similarly, a female gets one of her X chromosomes from her father whose development as a male it had determined. (See Table 6.) There is therefore a shuffling of X chromosomes from one sex to the other in succeeding generations, and it will be clear that the determination of sex must be regarded as quantitative rather than as qualitative. It must depend on the number of X-borne genes an individual possesses. If he has a single dosage he will develop as a male: if he has a double dosage he will develop as a female.

A still more important stage has recently been reached in the experimental work on this question. Just as we saw in Chapter II that most characteristics are controlled by groups of genes interacting rather than by one gene alone, so also the determination of sex has been shown to depend not merely on the possession of one or two X chromosomes, but on the *balance* between these sex chromosomes and the other chromosomes in the cell which are collectively known as *autosomes*. In some cases creatures have been found which possess an extra X chromosome—three instead of two. The female characteristics of these individuals are greatly exaggerated. Furthermore, it has been found possible by means of X rays to destroy portions of an X chromosome in a female, and it is then discovered that the larger the piece destroyed the more pronounced is the approach of that female towards maleness, until finally, if the whole of one X chromosome is destroyed, the individual develops as a normal male. So far, therefore, we may conclude from this that X chromosomes are female-determining.

Now, through accidents in cell division, some individuals

have occasionally been found which have three sets instead of two of *all* their chromosomes. These, instead of turning out to be extraordinary monstrosities, prove to be quite normal females. But if they have three sets of autosomes but only the ordinary pair of X chromosomes, instead of being normal females they turn out to be intermediate in sex. And when only *one* X chromosome exists in the presence of three of all the others the male characteristics are exaggerated. Again, the loss of portions of the autosomes produces a tendency towards the development of female characteristics. Consequently, we may conclude that whereas the X chromosomes are female determining, the autosomes are male determining.[1]

We have seen, therefore, that both the X chromosomes and the autosomes take part in determining sex. Which sex is produced depends on the balance between the number of X chromosomes and the number of autosomes. When the balance is tipped towards X chromosomes females are produced: when it is tipped towards autosomes males are produced, and according to the degree to which the balance is tipped different degrees of masculinity or femininity may occur. Either sex, therefore, is capable of taking on the characters of the other. Either can become intermediate or show various degrees of dosage of the other sex.

INTERSEXES

Apart from intersexuality caused by an abnormal balance between the X chromosomes and autosomes there are two other ways in which intersexuality has been found to occur. One of these, gynandromorphism, seems to be caused by an irregularity in an early division of the germ cell as the individual is developing. Sometimes, as a result of this strange cases occur in insects in which one-half of the creature is male and the other half female. This is particularly striking in those butterflies and moths in which the two sexes have different colourings. Thus in some butterflies one half of the body will show the brilliant red, gold or blue colours of the male, while the other half is the dull-brown, grey or white of the female. Not only are the external characteristics

[1] An apparent exception to this theory of the balance between sex chromosomes and autosomes as the prime determinant of sex arises among the hymenoptera. In the honey bee, for instance, eggs which are unfertilised and which therefore contain chromosomes derived exclusively from the female develop into males. Yet these eggs contain the female proportion of X chromosomes to autosomes (in this case one X chromosome to one set of autosomes). For a discussion of this point and for further references see Waddington (15) p. 229.

changed, but also the internal physiological ones. This curious phenomenon seems to be caused by a mishap in the first division of a fertilised egg bearing two X chromosomes—and potentially female—in which one X gets left behind and lost, so that while one of the cells contains two X chromosomes and is female, the other has only one and is male. Thus the respective sides of the body arising from these original cells will be female and male. In other cases instead of one half being male and the other half being female only a quarter or some other fraction is of one sex and three-quarters (or other appropriate fraction) of the other. This seems to be due to a mishap at a later stage of division than the first.

This type of intersexuality has not, however, ever been known to occur in mammals. In the insects in which it does occur it shows that the sex in every cell in the body is determined independently by its own chromosome outfit. In mammals, however, as in some other animals, the sex glands pass into the blood substances which circulate through the body and modify the rest of the sex characters. In these creatures, therefore, the chromosome balance determines not the sex of every cell, but whether male or female sex glands shall develop. Subsequent sexual development is handed over to the secretions from the sex glands.

In mammals the two sexes develop alike for the first few weeks after fertilisation, although they contain a different balance of X chromosomes and autosomes. The embryos develop in this "indifferent" condition up to and beyond the point where their heads, limbs and other organs are distinguishable. Then those which contain only one X, i.e. the males, begin to develop distinctive characteristics. Those which contain two X's, i.e. the females, remain in an indifferent condition until later on when they too take on distinctive characteristics. The fact that there is a period in development during which the male is developing distinctive characteristics while the female remains in an indifferent condition is of importance in connection with the second type of intersexuality—intersexuality in time rather than (as in gynandromorphism) in space—as we shall see in a moment.

When once the distinctive characteristics of the male and of the female have begun to show themselves the two sexes continue to develop differently right up to adult life. And these later sex differences are produced by different types of hormonal secretion from the testicular glands which have developed from the germ cells containing only one X and the ovaries which have developed from those containing two X's. The secretion from the testis produces the male secondary sex characteristics, e.g. beard, greater size, deeper voice, etc., while that from the ovaries produces the female secondary sex characteristics, e.g. mammary glands, different shape of pubic hair, etc. That this is so has been proved by experiments[1] in which the ovaries from young female rats have been removed and replaced by transplanting a testis taken from a young male, or by removing the testis from a young male and transplanting an ovary taken from a young female. In such cases the female body which has had a testis transplanted develops male secondary sex characters, while the male body containing a transplanted ovary develops female secondary sex characters and may even develop a mammary gland that produces milk and be able to suckle young.

A confirmation of these findings comes from an entirely different source. When twins of opposite sexes occur in cattle the male twin is a normal, healthy male but the female twin usually possesses both male and female characteristics: it is an intersex ("freemartin").[2] In these cases it has been shown that there has been some fusion between the blood circulation of the two twins, the blood from each twin passing into the body of the other, and the reason why the female twin is changed towards maleness while the male develops normally is probably because the male begins the development of his distinctive sex characters earlier than the female. The male can be identified as a male while the female is still in an indifferent state of development. As a result of this the hormonal secretion from the male is produced earlier than that from the female, and this, circulating in the body of the female, turns it in the direction of maleness. The result is that the female body, turned towards maleness, never develops her characteristic hormonal secretion and consequently the development of the male twin is not interfered with.

Sex Ratios

One final point calls for some comment. The XX, XY mechanism would offer a perfect mechanism for the numerical equality between the sexes. In fact, however, it is known that

[1] See, e.g. Jennings (6) pp. 79–80.

[2] This is the most usual situation, though sometimes the female is normal. In these cases there is no fusion of the blood streams of the twins.

among human beings at conception there is a considerable preponderance of males over females.[1] Throughout life that preponderance is reduced: more males than females die at all stages of development. At birth there are about 106 males to 100 females: equality is attained in late adolescence: over the age of 85 there are twice as many women as men. Part of this differential mortality is undoubtedly due to the greater risk of fatal accidents to which the male is exposed both in Western civilisation and in less advanced societies (though the type of risk is different in different communities), for though the female's greatest risk is at childbirth this is easily outweighed by the continuous risks to which the male is exposed in his working life, in wars and so on. But the rest of the differential mortality is probably due to a constitutional difference in vigour between the sexes. Males are less viable and more prone to illness than females. Males on the whole use up more energy than females—they have a higher "metabolic rate". It may be that their lower resistance is partly due to this greater expenditure of energy in addition to their greater exposure to accidents.

The fact that more male than female babies are conceived naturally raises the question why and how this should be so. One theory is that since the Y chromosome is so much smaller than the X those sperms which contain the Y chromosome will be smaller and lighter than those containing the X. Their mobility will therefore be greater, and they will have a better chance of fertilising the egg than those containing the X chromosome. This theory of relative mobility, however, is not always confirmed when one examines the relative proportion of males to females born in other kinds of animal. In the pigeon, for example, there are 115 males born to every 100 females,[2] and in the pigeon, as in other birds, it is the female sex which is heterogametic (i.e. which contains the XY chromosome) while the male is homogametic (XX). All the sperms are of the same chromosomal character, and it is the eggs which differ and determine the sex of the offspring. There is no obvious reason, therefore, why some sperms should travel more rapidly than others, and even if they did this would not affect the sex of the offspring.

The same difficulty arises in connection with other views that have been put forward, e.g. that one of the two types of sperm

[1] This has been studied in detail only among peoples of Western civilisation, but is probably true for the whole human race.

[2] See Jennings (6) p. 84. Data from Goldschmidt, R. B., *The Mechanism and Physiology of Sex Determination*, 1923.

in mammals is more resistant than the other with the result that certain unfavourable environmental conditions would destroy one type more easily than the other, so that eggs would tend to be fertilised by the uninjured type. These unfavourable environmental conditions might be due to chemical conditions or temperature or types of food affecting the acid-alkali balance in metabolism. Or again the eggs might have certain physical or chemical properties so that they admitted one type of sperm more readily than the other, and these properties might be alterable by the conditions of the environment of the egg. Nevertheless, since in man the male is the heterogametic sex and the sperms are consequently of two types all these factors must be borne in mind as possible causes of the difference in numbers of boys and girls conceived. If any of them were found to be important it might be possible to account for and to influence the fact that some families appear to produce through several generations a higher proportion of children of one sex than of the other.

Finally it is known that in some organisms the sex of the individual depends on the conditions in which it develops, e.g. "(a) In a certain common mollusc called Crepidula plana, if the eggs are kept away from full-grown individuals, they develop into females. But if they are allowed to develop near to older specimens of Crepidula, they become males. In such eggs there is thus a very delicately balanced condition, which may be turned toward either sex by a slight change of conditions.

"(b) In a marine worm known as Bonellia, the female is a large worm with a proboscis. The male, on the other hand, is very small, and is parasitic within the body of the female (in the uterus). The eggs develop into small creatures that swim about in the water. If they find a female, they attach themselves to her proboscis and develop into males. But if they do not become attached to a female, they themselves develop into females.

"Here again there is a delicate balance as to sex: which direction development shall take depends on the external conditions."[1]

Physiological and Mental Differences

Apart from these fundamental biological similarities and differences between the sexes the cultural environment almost

[1] Jennings (6) pp. 86–7.

certainly plays some part in determining many of the other physical sexual differences between men and women. A possible exception to this is in the attainment of maturity. Girls reach maturity earlier than boys in physical characteristics. At each age between 6 and 17 girls attain a higher percentage of their adult height and weight than boys.[1]

	Height		*Weight*	
Age	*Boys*	*Girls*	*Boys*	*Girls*
17.5	100	100		
16.5	97.5	99.2	100	100
15.5	94.5	98.3	88.7	95.1
14.5	90.3	96.3	78.9	87.4
13.5	86.4	93.3	70.0	79.0
12.5	83.4	89.4	63.5	70.0
11.5	80.6	85.6	58.4	61.8
10.5	78.0	82.5	54.1	56.0
9.5	75.1	79.3	49.0	51.0
8.5	73.3	76.1	45.0	46.7
7.5	69.1	72.8	40.9	42.4
6.5	65.9	69.0	37.4	38.5

TABLE 7. Percentage of final growth which has been attained at ages preceding maturity. (From Anastasi (3) p. 397.) (Data from Lincoln, E.A. (7) p.19.)

In emotional maturity, too, girls appear to be more advanced than boys—though this is less likely to be wholly hereditarily determined.

Boys are better than girls in skills requiring a gross co-ordination of muscles, but girls are better than boys in skills requiring fine co-ordination. Much of this may be due to the socially determined play and leisure activities of the sexes. The skill in fine co-ordination of the girls, for example, may be developed by practice in sewing and embroidery, that of the boys in gross co-ordination by their more active pastimes. Similarly, too, any original and innate differences in muscular development that there may be between the sexes are fostered and magnified by the difference in their activities which our culture regards as suitable.

In menstruation, pregnancy and childbirth we have definite physiological facts which are not shared equally by the two sexes. Investigations have shown that during menstruation a woman's

[1] See Anastasi (3) pp. 397–9.

efficiency is reduced,[1] and its incapacitating effect is a common experience of everyday life. There are, however, grounds for believing that even here cultural influences are important in determining the extent of the incapacitation. There are grounds for believing that at the present day, and particularly perhaps in America, the degree of incapacitation is far less than it was during the latter half of the past century. But women have two other major handicaps from which men do not suffer, namely pregnancy and childbirth. The disturbance of metabolic processes during pregnancy often causes unusual food preferences and other physiological symptoms. It is not unreasonable to suppose that such changes have their psychological concomitants. Nevertheless, it has been shown that except for a relatively short time both before and after childbirth most women are able to continue their jobs satisfactorily. If they do not do so, may it not be that they do not choose to do so? Attention directed on to the care of a child may be attention diverted from interest in a job. It is easy to show that, as our society is organised, there is no true alternative open to women. They *have* to look after their children.

The reason these points are raised is because they lead us directly to a consideration of other psychological and sociological differences between the sexes. One of the most important of these problems is the question whether men possess more natural ability than women. Far more men than women have achieved positions of eminence, and to account for this fact some people believe that men are more highly gifted than women. Although we cannot answer this point categorically, there is no evidence to suggest that intelligence (or whatever the qualities may be that make for eminence in our civilization) is a sex-linked characteristic, a gene in the Y chromosome that therefore only appears in males, or even a sort of catalytic agent in the Y chromosome that sets off or makes more effective something that already exists in the X. What evidence there is, in fact, suggests that there is no substantial difference in the abilities of men and women.

In the first place when intelligence tests which have been standardised on a combined group of males and females are given to men and women separately, it is found that the difference in their average scores is negligible. The differences in scores obtained by individuals *within* either group are of far greater importance than the difference in scores *between* the groups.

[1] See McCance, Luff and Widdowson (9). Also Sowton, Myers and Bedale (13).

Girls, in fact, on the whole tend to get a higher score than boys, particularly in the lower age groups and when the test used to determine their intelligence is largely based on words and language. On the practical kinds of test, however, boys usually score more highly than girls. So far as tests are a reasonable measure of ability, therefore, there is no good reason to account for the difference in eminence between men and women.[1]

It might be, however, that though the averages were the same the scatter in a group of men was wider than in a group of women. The same average may be obtained from the scores of two groups when in one group the individual scores are closely clustered about their mean while in the other there is a wide dispersion of the scores. But if this were so in the case of women's and men's intelligence, then we should expect to find a greater number of men than of women not only in positions of eminence, but also in institutions for mental defectives. In fact, however, the difference is negligible. On 1 January, 1933, for example, there were 33,344 males and 32,549 females under care in institutions for mental defectives in England and Wales—that is, 795 more men.[2] The population of England and Wales in 1931 was 19,133,000 males and 20,819,000 females, and on these large figures a difference of 795 is insignificant. The difference is made still more insignificant by the fact that there is a greater likelihood that feeble-mindedness amongst men will be diagnosed more readily than it will among women. Women in peace time undertake much more sheltered occupations than men—occupations like domestic service and housewifery—and in these feeble-mindedness is less readily detectable than it will be in most of the occupations that are open to men.

In so far as women have achieved eminence it has usually been in some occupation like literature in which the cultural prejudice against them has not been so strongly marked. It is reasonable to argue from this that were cultural prejudices to be further relaxed many more women would achieve eminence in other spheres. Society expects more of men than it does of women. It expects men to struggle and strive, to be aggressive and to do better than other men. The training of the majority of boys in western civilisation is directed from their earliest years towards making them "tough", and parents are often afraid that if they

[1] Anastasi (3) pp. 425–8.

[2] 19th Annual Report of the Board of Control (12) p. 74. More recent Reports show a very similar picture.

show their boys too much affection they will turn out to be weak and ineffective creatures. Girls, on the other hand, tend to be sheltered and protected by home influences. Furthermore, the education given to boys gives them an advantage later on when they seek a job.

An indication of the influence of cultural forces is seen in the results of Terman and Miles' M-F test.[1] By carefully choosing from a large number of items which serve to indicate a person's predominant interests and attitudes, a significant difference in average scores was found between groups of men and of women. There was, however, a good deal of overlapping, showing that some men possess interests which are more feminine than those of the average woman and *vice versa*. Within the groups of men and women the relative masculinity or femininity of the scores seems to depend to a large extent on the kind of group to which the person belongs. Thus, athletes, or men who have acquired some outstanding position in their careers, score more highly on the masculine side than theological students: and passive male homosexuals get about the same feminine score as the average woman. Similarly, among the women those with the most feminine scores are those whose main interests are domestic. Such differences in interests may turn in part on innate physical characteristics—the athletic men, for example, being likely to be already above the average in physique before they become athletes. But it is unlikely that this is sufficient to explain the whole difference within groups of men or of women, or between the sexes.

Consistent sex differences have also been found in the Allport-Vernon "Study of Values"[2] which shows that the highest average values for men lie among the theoretical, economic and political interests, while for women they lie among the aesthetic, social and religious interests.

Furthermore, the type of interests revealed in conversation has been the subject of study in both New York and London.[3] In both New York and London when women were having a conversation with another woman, the principal topics of conversa-

[1] Terman and Miles (14).
[2] Allport and Vernon (2).
[3] Murphy, Murphy and Newcomb (11) pp. 693–4. It should be noted, as Professor Pear points out to me, that this experiment was performed on a very restricted sample, and that it is unjustifiable to assume without further evidence that the results are applicable generally to all conversations between men and women, whatever their social class or whatever part of the country they come from. The example is included here solely for purposes of illustration.

tion were clothes and social affairs. When men were having a conversation with another man the main topics were money and business affairs. There was a difference between the two countries, however, when conversations were occurring in which the participants were of different sexes. Instead of getting a mixture of the two sets of topics it was found that in the United States women tended to adapt themselves to the topics of chief interest to men: in England, however, it was found that men tended to adapt their conversation to the topics of chief interest to women. The influence of cultural factors probably accounts for this difference.

Sociological Causes of Sex Differences

If one looks at the anthropological literature one sees that all societies have prescribed different types of behaviour for men and women, though whether any particular type of behaviour is ascribed to a man or to a woman has varied in different communities. Linton[1] points out that even the psychological characteristics attributed to men and to women vary so much in different societies that it is difficult to believe that they can have a physiological basis. Thus, "Our own idea of women as ministering angels contrasts sharply with the ingenuity of women as torturers among the Iroquois and the sadistic delight they took in the process. Even the last two generations have seen a sharp change in the psychological patterns for women in our own society. The delicate, fainting lady of the middle eighteen-hundreds is as extinct as the dodo."[2]

The same thing is true of the occupations which are assigned to men and to women. Among the Arapesh women carry heavier loads than men because their heads are thought to be so much harder and stronger. "In some societies women do most of the manual labour; in others, as in the Marquesas, even cooking, housekeeping, and baby-tending are proper male occupations, and women spend most of their time primping. Even the general rule that women's handicap through pregnancy and nursing indicates the more active occupations as male and the less active ones as female has many exceptions. Thus among the Tasmanians seal-hunting was women's work. They swam out to the seal rocks, stalked the animals, and clubbed them. Tasmanian women also hunted oppossums, which required the climbing of large trees."[3]

[1] Linton (8). [2] Linton (8) p. 116. [3] Linton (8) p. 117.

Margaret Mead[1] has made out a determined case for the influence of the cultural pattern of society on sexual differences in behaviour. She extends the theory developed by Ruth Benedict[2] (that a society's cultural pattern is one of the most important determinants of the behaviour of its individual members) to cover the differences in the way the sexes behave. It is the pattern of culture rather than genetical differences that chiefly determine characteristic masculine and feminine modes of reaction. Mead describes three tribes. The first, the Arapesh, is a gentle, peace-loving people amongst whom there is a minimum of distinction between men and women. Boys and girls play the same games which are mild and co-operative and never competitive. The child is brought up as much by the father as by the mother, and its development is lovingly fostered with much petting and fondling. There is no acknowledged temperamental difference between the sexes in adulthood, and a sex relationship is as likely to be initiated by a woman as by a man. Sex is a serious business, and there are few liaisons or casual encounters. Those who are maladjusted in this tribe are the violent, aggressive men and the violent, aggressive women. This offers an immediate contrast with our own society in which it is the unaggressive men and the agressive women who tend to be the deviants.

The second tribe Mead describes, the Mundugumor, also assumes that men and women have similar temperamental endowments, but whereas in the first tribe they are both regarded as gentle and unaggressively sexed, in this tribe they are both expected to be violent and competitive and aggressively sexed. The children are born into a hostile world. Both father and mother resent the birth. The children are suckled with great haste and returned to their baskets as soon as they stop sucking, even for a moment. They therefore hurriedly gulp down what they can while they can, and in the process they frequently choke. This infuriates them as well as angering their mother. The suckling situation is characterised by a contest and a struggle instead of by affection and reassurance such as is found in the first tribe. This kind of upbringing has its effect on the behaviour of the sexes as they grow up. The love affairs of the young unmarried people are characterised by violence and passion rather than by tenderness or romance. Those who deviate from this pattern are the unaggressive, gentle men and women. It is the same type for

[1] Mead (10). [2] Benedict (4).

both sexes, as it is in the first tribe, but the type itself is very different.

Finally Mead describes a third tribe, the Tchambuli. Here it is the women who have the real power and who own the property and who do all the important work. The men have to wheedle what they want out of women by means of languishing looks and soft words. The women do the fishing and the weaving; the men go about dressed up in fine ornaments and arrange dances and ceremonies. The women are in the superior position. "The women's attitude towards the men is one of kindly tolerance and appreciation. They enjoy the games that the men play; they particularly enjoy the theatricals that the men put on for their benefit." And whereas the lives of the men are one mass of petty bickerings, misunderstandings, reconciliations, avowals, disclaimers and protestations accompanied by gifts, the lives of the women are singularly unclouded with personalities or with quarrelling. For fifty quarrels among the men there is hardly one among the women. "Tchambuli women work in blocks a dozen of them together, plaiting the great mosquito bags from the sale of which most of...[their economic existence is]...obtained. They cook together for a feast, their clay fireplaces . . . set side by side. Each dwelling house contains some dozen to two dozen fire places, so that no woman need cook in a corner alone. The whole emphasis is upon comradeship, efficient happy work enlivened by continuous brisk banter and chatter. But in a group of men there is always strain, watchfulness, a catty remark here, a *double entendre* there. 'What did he mean by sitting down on the opposite side of the men's house when he saw you upon this side?' 'Did you see Koshalan go by with a flower in his hair? What do you suppose he is up to?' "[1]

The men sit about in a highly-charged atmosphere of courtship, none knowing upon whom a woman's choice will fall. Each youth holds his breath and hopes, and no young man is willing to trust another. It is women who are supposed to be the more highly sexed: the men are much more easily able to wait.

In the Tchambuli, therefore, we see a reversal of the social expectation found in our own society. The more virile, agressive and actively-sexed males, and the gentle, maternal females tend to be the deviants.

[1] Mead (10) p. 253.

Psychological Reasons for the Differences

Mead, therefore, presents a sociological theory which attempts to express the behaviour of men and women in terms of a norm determined by a pattern of culture. A more purely psychological theory which endeavours to explain the behaviour of women in western civilisation has been devised by Alfred Adler.[1] Adler holds the view that everybody possesses some feeling of inferiority in relation to the world around him, and that this feeling plays an important part in determining his behaviour. The feeling arises in childhood when the child naturally feels weak in relation to the outside world. His actions are determined for him by his parents who often make him do many things he would rather not do, and prevent him from doing others which he wants to do. The same feeling persists after the child has grown up. There, too, he feels insignificant and impotent in the face of society, its traditions and conventions. Now some children and adults feel this sense of inferiority more strongly than others. It is particularly likely to be strong in someone who suffers from a physical handicap, such as low stature, a club foot, a cleft palate, and so on. And the more strongly this sense of inferiority is felt, the stronger will be the attempt to compensate for it by trying to demonstrate his superiority in some other way, and by setting himself an ambitious goal to achieve. So long as he possesses the necessary ability he will manage to achieve his goal, but many people are not so lucky: in spite of their acute inferiority feelings they do not have the ability to achieve the high goals they set themselves as a result of this feeling. These become neurotics.

Apart from physical infirmities there are various other factors which determine the inferiority feeling and its resulting compensations. One is the ordinal position of the individual in his family. Adler maintains (without much evidence to support the view, it is true) that younger or youngest children are more likely to achieve eminence than older children, for older children have at least the younger children over whom they can feel superior: the younger children have no one below them. The younger children will therefore tend to set their goals higher than the older children in compensation for this fact, and consequently, if they possess the necessary ability, they will be more likely to achieve eminence. It should also follow from this theory that a larger number of younger than of older children will be found among the neurotics, namely those who compensate more strongly

[1] Adler (1) pp. 120–48.

than the older children for their inferiority feelings but do not possess the necessary ability to achieve their goals—but there is no evidence whatever to support this.

Now, so far as women are concerned Adler has developed his theory of the "masculine protest". Our social organisation is a chiefly masculine affair, mainly run by men for men. Masculine prestige is very much higher than feminine. From early childhood this state of affairs exists. Girls take every opportunity they can to play boys' games, to wear boys' clothes, to ride boys' bicycles, etc., whereas boys show much resistance to indulging in girlish pursuits. Later on this is reinforced by the emotional connotation surrounding language. Words like "masculine", "manly" and "virile" have a pleasant, those like "effeminate" an unpleasant connotation.[1] This being so, one would expect the same principles to apply to women as are held to apply to younger children. One would expect more women than men to be neurotic, which happens to be true, but one would also expect more women than men to achieve positions of eminence, which is most decidedly untrue—unless there is a psychological difference in their make-up, so that women do not react in the same way as younger boys. Perhaps Adler implies that there is, for although he claims that some women do attempt to domineer (militant suffragism being used as an example of the attempt on the part of some women to demonstrate that they are *superior* to men), yet other mechanisms are more often employed. One of the most frequent, Adler claims, is for women to pretend to accept a position of inferiority, and in consequence to throw all the burden of responsibility on to men. Then, having left the men to make all the important decisions, these women will be in the happy position of being able to point out all the disadvantages of those decisions, and all the advantages of a different decision—whatever the decision may happen to have been. It would thus seem to follow from Adler's theory that the type of superiority which women on the whole tend to achieve is not of the active, glamorous and resplendent type—they leave that to men—but much more of the subtle and self-effacing type. Their leadership is like that of some children who keep themselves in the background and play an apparently unimportant role in group games, but

[1] It should be pointed out, however, in criticism of Adler's theory that "effeminate" is derogatory not because of its connection with femininity but because of its usual application to males who are deficient in masculine characteristics. The adjective "mannish" applied to women is correspondingly derogatory whereas the adjective "feminine" is not.

really control and direct the other children from their positions in the background. Women, therefore, may really be just as eminent as men, but their eminence is of a type that it is less blatant and obvious.

Conclusion

We have seen therefore that the genetic aspects of sex depend in the first place on the balance between X chromosomes and autosomes. As the relationship between them alters different degrees of masculinity and femininity may appear. Apart from the sexuality that is dependent on chromosomal balance we find in mammals and in some other organisms that the development of other sexual characteristics will depend on hormonal secretion from the genital glands. At this stage too, therefore, there is a possibility of different degrees of masculinity or femininity appearing as the result of a stronger or weaker degree of secretion. Thus a more or less masculine male or a more or less feminine female may be the result of either or both of these factors—chromosomal balance and glandular activity. The cumulative physiological effect of the differential glandular secretion which persists from the first weeks of life to the climacteric or beyond may clearly be very considerable, and it may have important sociological and psychological repercussions, both when the amount of secretion lies within the normal limits and also when it diverges sharply either above or below the normal. This, coupled with the physiological disadvantages of menstruation, pregnancy and childbirth from which women suffer reinforces the differences which men and women may show in their behaviour.

One other physiological factor may be of importance in the differences between men and women. The male's metabolic rate is higher than that of the female. The male uses up his energy faster: his life is shorter and he may thus concentrate a greater fund of energy into this relatively shorter life than the female is able to do. It is conceivable that this greater and more concentrated expenditure of energy plays some part, even though possibly a small part, in the greater degree of eminence which men have attained in Western civilisation.

On the other hand there is evidence, even from the lower organisms, that the kind of environment in which the organism is reared has a considerable effect on the development of male or female characteristics, and when one turns to human beings one finds that the behaviour of men and women is very strongly influenced—some say even principally determined—by the sociological and psychological influences of the culture in which they are brought up. Such tests as can be employed do not for the most part indicate any startling differences in men's and women's abilities. And those in which reliable differences have been shown to exist between the average score of men and of women are precisely those, like the Masculinity-Femininity tests, which are most strongly influenced by environmental factors. Furthermore Mead has described a tribe in which the rôles played by men and women in Western civilisation appear to be reversed—the men showing the kind of characteristics which in our society are usually attributed to women and *vice versa*. Finally Adler has stressed from the psychological point of view the implications of the male prestige which exists in Western communities, and though the deductions which one may make if one carries his premises to their logical conclusion do not all turn out to be true, Adler has played an important part in focusing attention on the influence which psychological factors, dependent on a bolstered-up male superiority, may sometimes have on the relative behaviour of men and of women.

Thus we see again in connection with sex differences, as we saw in chapters II and III in connection with other differences, that similar ways of behaving may be produced by different causes. Sometimes a typically masculine or feminine type of behaviour may depend on chromosomal balance or be the consequence of a certain degree of hormonal secretion: in other cases a very similar form of behaviour is the result of psychological reactions to environmental forces dependent on the cultural pattern in which an individual happens to be reared. In sexual characteristics as in other characteristics what heredity can do environment can also do.

REFERENCES

1. Adler, A., *Understanding Human Nature.* London: Allen & Unwin, 1928, pp. 286.
2. Allport, G. W., and Vernon, P. E., "A Test for Personal Values", *J. Abn. Soc. Psychol.*, 1931, *26*, 231-48.
3. Anastasi, A., *Differential Psychology.* New York: Macmillan, 1937, pp. 615.
4. Benedict, R., *Patterns of Culture.* London: Routledge, 1935, pp. 291.
5. Ford, E. B., *The Study of Heredity.* Home University Library, London; Thornton Butterworth, 1938, pp. 256.
6. Jennings, H. S., *Genetics.* London: Faber, 1935, pp. 351.

6A. KLEIN, VIOLA, *The Feminine Character: The History of an Ideology.* International Library of Sociology and Social Reconstruction. London: Kegan Paul, 1946.
7. LINCOLN, E. A., *Sex differences in the Growth of American Schoolchildren.* Baltimore: Warwick and York, 1927, pp. 189.
8. LINTON, R., *The Study of Man.* New York: Appleton Century, 1936, pp. 503.
9. MCCANCE, R. A., LUFF, M. C., AND WIDDOWSON, E. E., "Physical and Emotional Periodicity in Women", *J. Hyg.*, 1937, *37*, pp. 571-611.
10. MEAD, M., *Sex and Temperament in Three Primitive Societies.* London: Routledge, 1935, pp. 291.
11. MURPHY, G., MURPHY, L. B., AND NEWCOMB, T. M., *Experimental Social Psychology.* New York: Harper, Rev. Edn., 1937, pp. 1121.
12. *Nineteenth Annual Report of the Board of Control for 1932.* London: H.M. Stationery Office, 1933, Part I.
13. SOWTON, S. C. M., MYERS, C. S., AND BEDALE, E. M., "Two contributions to the experimental study of the menstrual cycle", *Ind. Hlth. Res. Bd. Rep. No. 45.* London: H.M. Stationery Office, 1928, pp. 72.
14. TERMAN, L. M., AND MILES, C. C., *Sex and Personality.* New York: McGraw Hill, 1936, pp. 600.
15. WADDINGTON, C. H., *An Introduction to Modern Genetics.* London: Allen & Unwin, 1939, pp. 441.

CHAPTER V

RACE DIFFERENCES

"However, when they had been running half an hour or so, and were quite dry again, the Dodo suddenly called out 'The race is over" and they all crowded round it, panting, and asking 'But who has won?' . . . At last the Dodo said '*Everybody* has won, and *all* must have prizes.' "—*Alice in Wonderland.*

Introduction

Ginsberg[1] has defined a race as "a group of people who, within given limits of variation, possess in common a combination of hereditary traits sufficient to mark them off from other groups." The criteria for race, according to this definition, and indeed according to most definitions, are therefore physical (whereas the criteria for distinguishing between nations are largely psychological). But difficulties in classification at once arise in determining what "limits of variation" we shall allow ourselves. Whether or not a particular group of people may be referred to as belonging to a particular race will largely turn on the degree of variation of these hereditary traits which we allow ourselves to employ.

By stringent selection combined with inbreeding for several generations (say seven or eight) it is possible to produce a "pure line", that is to say a group of organisms which will have the same genetic constitution—in so far as the selection has been applied—and which will breed true except for occasional mutations. So far as human beings are concerned this "ideal" has never been achieved. The nearest approach to it, perhaps, occurs in those few groups which for geographical or social reasons have been isolated from other groups. Such, for example, are the inhabitants of Pitcairn Island where the descendants of the six English mutineers of the *Bounty* and the Tahitian women who accompanied them have lived and inbred since 1790; or among the Esquimaux or in Melanesia where contact with others has been negligible; or among the Bastaards in South Africa, or some Anglo-Indian mixtures in India where the isolation has been social rather than geographical. The Bastaards are descendants of Boers and Hottentots: they have kept themselves as an inbreeding group apart from the whites or the natives, and in the course

[1] Ginsberg (16) p. 56.

of several generations they have developed a stable type which is distinct from either of the groups from which they are descended. The same thing is reported to be happening to-day among Anglo-Indian groups in India, even in the towns.[1]

But in such cases the selection (in the sense of eliminating those who possess characteristics which diverge from those of the rest of the group) has never been sufficiently stringent to enable a pure line to develop, and similarly such mutations as may have occurred have not been weeded out. Even here, therefore, it is unjustifiable to speak of these peoples as forming races in the strictest sense.

How much more then is it unjustifiable to speak of other groups as forming races with similar hereditary characteristics? Nevertheless many attempts have been made to classify human beings into races. The scientifically trained person feels an almost irresistible tendency to classify his specimens. And, besides, a classification has sometimes proved useful for political purposes.

The difficulties are considerable. In the first place all the races of mankind can interbreed and produce fertile offspring. In the second place there is no justification for the view that some races are "prepotent" over others, e.g. that in crosses between "primitive" and more advanced races the "savage" type will invariably break through. No one race shows a predominance of all its characteristics over another: all that may happen is that if one group possesses more dominant genes than another then in the crosses between the two stocks the dominant characteristics will be more conspicuous. That is why, as Baur, Fischer and Lenz[2] point out, when Negroes and Europeans are crossed the resulting offspring are most often dark skinned and dark haired. The view that the savage type breaks through is largely due to a neglect of negative instances. Thus in crosses between Jews and gentiles the tendency is to neglect those instances in which only non-Jewish characters are obvious and to concentrate attention on those in which black hair, convex nose, etc., appear. Yet Baur, Fischer and Lenz[3] maintain that in crosses between broad-nosed Jews of the Eastern stocks (Ashkenazim) and North Europeans with a sharply-cut, narrow nose, this Nordic nose is manifestly dominant.

All races are heterozygous for a very large number of factors.

[1] Benedict (4) p. 41. [2] Baur, Fischer and Lenz (3) p. 175.
[3] Baur, Fischer and Lenz (3) p. 176.

The most one can do, if one insists on trying to classify them, is to pick out the most distinctive characters and to neglect those that are less conspicuous. Thus if the number of characters used as distinguishing marks is sufficiently strictly limited a few large groups of mankind will emerge. This is a dangerous procedure because it means using phenotypes as criteria for classifying genotypes, and this, as we saw in Chapter II, may be extremely inaccurate. In addition to this, as we shall see in a moment, it often turns out to be quite arbitrary.

Physiological Distinguishing Marks

One of the most obvious and most frequently employed divisions is into differences of skin colour—white, black, yellow, red and possibly brown. But by itself skin colour is an insufficient racial determinant. The same pigments are found in all human skins, and the resulting colour depends on the *degree* of pigmentation present[1] which in turn partly depends on environmental conditions, such as exposure to the sun's rays. Then there is a continuous range of variation from the lighter to the darker shades, and there are no clear-cut lines of transition. Furthermore, some groups of whites are darker than some negroes,[2] and in many there is a considerable degree of overlapping so that the darkest white is darker than the fairest negro. Again, if it is used by itself its results will conflict with the evidence drawn from the use of other criteria, e.g. using skin colour as the criterion Australian aborigines will be grouped with negroes.

Eye colour and eye form are undoubtedly hereditary and they offer a sharper contrast between types than is true of skin colour. To this extent they are superior as potential determinants of race. Unfortunately, however, there are not a sufficient number of different hereditary shades of eye colour to be of much use. Black eyes are common in all groups: in fact they are practically universal outside a single group—although within this group they have been used (unsatisfactorily as we shall see) as a criterion for determining exclusion from the "Nordic race".

Eye form, in particular the "slanting eye" caused by a fold of skin (the epicanthic fold) covering the inner angle of the eye, is typical of yellow-skinned Asiatics. It is not, however, exclusively confined to them. It occurs among some negroes and also in infancy among many whites, though it does not usually persist in them after childhood.

[1] Anastasi (2) p. 455. [2] Benedict (4) p. 23.

There is a continuous series of gradations in hair colour from the darkest to the lightest, and as dark hair, like dark eyes, is found in every group, hair colour does not make a satisfactory criterion. Fair hair, like blue eyes, has been used as a criterion for determining membership of the "Nordic race", but it is quite as unsatisfactory.

So far as hair form is concerned there are three main types—straight lank hair, wavy hair, and frizzy hair. But, as with skin colour, their use sometimes cuts across races described in terms of other criteria. For instance, if skin colour is used as a criterion, the Australian aborigines will be classed with negroes: if, however, hair form is used then they will be classed with Europeans.

Stature is clearly hereditary and it may, perhaps, be usefully employed in differentiating pygmies from other groups. But so far as the other groups are concerned there is, as with many of the other criteria, a marked degree of overlapping between the groups. To say that the Lapps are short and that the Swedes are tall does not mean that the tallest Lapp is shorter than the shortest Swede. It only means that the average heights and the range of variation are different. And where there is a fair degree of overlapping between the groups the criterion loses a great deal of its usefulness. It will at best be limited to a statement about the groups as a whole: it will not be possible to apply it to any specific individual. The same difficulty holds true of other continuously varying characteristics, e.g. hair colour or pigmentation. Furthermore, to use stature as a criterion of race cuts across other classifications, e.g. both the tallest and the shortest groups which have been measured are negro, and on the island of Tierra del Fuego one of the tallest American Indian tribes is the near neighbour of a very short tribe.

The cephalic index is the relationship between the breadth and the length of the head. It is obtained from the formula:—

$$\text{Cephalic index} = \frac{\text{Breadth of head} \times 100}{\text{Length of head}}$$

If the index is 75 or less the individual is usually regarded as long-headed (dolichocephalic): if it is 80 or more he is regarded as round-headed (brachycephalic). These figures are more or less arbitrary, as are the dividing lines between the groups of other continuously varying characters, and for this reason, if for no other, they are unsatisfactory for use as criteria for racial differentiation. Again, there is not complete unanimity about the dividing line: some authorities believe that 77 and 81 are more appropriate limits. But there are other reasons why it is not a satisfactory criterion. It too conflicts with other measures. It does not distinguish the whites from the yellows nor from the negroes. Furthermore, some American Indians have the narrowest skulls that have ever been measured, while others have the broadest. Its main use has been to distinguish between subgroups within a larger group, e.g. between Nordics and Alpines within the whites. But obviously in so far as its specificity is increased its general usefulness is diminished.

When we turn to blood groups we turn from phenotypic to genotypic characters. Bloods belong to four groups, AB, A, B, and O. The heredity of these groups is known, and they might therefore offer a useful criterion for distinguishing between races. For instance, if both parents belong to group O, the child must also belong to group O. If the child has blood of type A, then one or both of his parents must be of type A, and so on. Therefore, when different blood types are found in any group one can be fairly certain that that group has had a mixed ancestry.[1] But again, if blood groups are used, the "races" that emerge are quite different from those which emerge from other types of measurement. Nordics, Alpines and Mediterraneans, for example, do not differ significantly in the frequencies of their different blood groups, the Japanese show a very similar picture to the Germans, and the Poles to the Chinese. The picture presented by the English is similar to that given by Sicilians, Tyrolese and Berbers.[2]

Now, with all this diversity of unsatisfactory criteria for distinguishing between races how is it possible to make any scientifically satisfactory grouping? Since most of the criteria are quantitative variables, it is somewhat arbitrary where the differentiating points shall be drawn, and, in addition, a classification according to any one of them cuts across a classification according to another. Nevertheless, let us see where some of them will lead us.

Classifying by skin colour leads to a division into white, yellow, black, red and brown: classifying by hair form leads to a division into three main groups—the straight, lank-haired group which includes Mongolians, Esquimaux and American

[1] Unless it is assumed that the same blood groups have arisen in different communities by mutation.

[2] Klineberg (25) p. 43.

Indians; the wavy-haired group which includes Europeans, Sudanese, Southern Indian and Australian; and the frizzy-haired group which includes Hottentots, African Negroes and Melanesians: classifying by blood groups gives some of the unexpected results we have already mentioned which differ from the other two classifications: and classifying by cephalic indices gives two groups only—the dolichocephalic Eurafricans and the brachycephalic Eurasiatics.

Nevertheless many anthropologists who are fully aware of the arbitrariness of their measures continue to find it convenient to make a classification of races, and to describe physical characteristics which are typical for each group. One such classification[1] is the following:—

I. *Negroid* (Negro, Melanesian and Hottentot). Frizzy hair; relatively little hair on body and face; long arms and legs; full lips; flat nose; black or dark brown in colour.

II. *Mongoloid* (Mongolian, Malay, American Indian). Straight hair; very little hair on face and body; short arms and legs; thin lips; often the epicanthic fold; yellow or reddish-brown in colour.

III. A group of doubtful classification which shows the characteristics of more than one of the other groups. These include the Australians and the Polynesians.

IV. *Caucasian or White* (Nordic, Alpine, Mediterranean and Hindu). This group has been studied in most detail. The Nordics are said to be tall, fair-haired, blue-eyed, fair-skinned and dolichocephalic: this is a type covering Northern Europe horizontally and including Scandinavia, England, Scotland, Northern France, Holland, and Northern Germany. The Alpines, or Central Europeans, are medium-sized, rather darker than the Nordics, and definitely brachycephalic: they are found in Central France, Switzerland, Northern Italy, Southern Germany, Austria, Poland, the Balkans, Russia and parts of Western and Central Asia. The Mediterraneans, or Southern European type, is short, has black or brown hair and eyes, relatively dark skin, and is dolichocephalic: it is found in Portugal, Spain, Southern France, Southern Italy, Greece, North Africa and Arabia.

The Hindu bears a close resemblance to the Mediterranean type.

But how far can such a classification take us? Let us consider that supposedly most Nordic of the Nordic group, the Swedes. If the student includes in his group only those Swedes who are

[1] cf. for example, Kroeber (27) pp. 34–57.

tall, fair-haired, blue-eyed, fair-skinned and long-headed he will find that he does not get 100 per cent. of Swedes conforming to this type, nor 80 per cent., nor 50 per cent., nor 25 per cent., nor 15 per cent., but only 11 per cent.![1] Would any botanist or zoologist think it worth while to define a group in terms of five criteria when in fact only 11 per cent. of his specimens agreed in possessing all those five criteria together? It means either that the criteria are too variable to be definitive or that his population has been so swamped by interbreeding with other groups that the original characters are no longer typical of the group. And the same thing is true wherever the anthropologist looks, with the exception of the few small isolated communities which have inbred for several generations and which we have mentioned at the beginning of this chapter—and these groups are insignificant in comparison with the rest of the world's population. The ideal type turns out to be as difficult to find as the rainbow's end. Each of the characteristics he hopes to be definitive may be found in existence in several other groups—though the frequency with which they appear in the different groups may be different. If he confines himself to relative frequencies a classification of human beings into groups may be useful. The danger is that the next and illicit step will be taken, and the frequencies which are valid for distinguishing between groups will be applied to determine to which particular group a *particular individual* belongs. The political temptation to do this is sometimes too strong to be resisted. But there is a world of difference between the two procedures.

Theories of Racial Superiority and Inferiority

Nevertheless, that step has been taken, and with it is associated the desire to prove that one or other group is "superior" to the rest or that one is more primitive than another. Aristotle in his *Politics* argued that the intellectual leadership of the Greeks followed of necessity from their favourable geographical position midway between the cold parts of Northern Europe which developed courage and good physique but not intellect, and the warm regions of Asia which developed intellectual prowess but sapped energy. Other groups have made the same claim, and often based their claims on similar arguments—that they alone of all groups occupy a favourable position midway between two extremes.

[1] Retzius and Fürst, quoted by Benedict (4) p. 29.

In more recent times a number of arguments have been put forward to show the superiority of the white race over all others. The whites are regarded as having advanced further along the evolutionary tree from their primeval ancestors than either the negroids or the mongoloids. In support of this theory points of greater similarity are often drawn between the negroids and anthropoids. It is mentioned, for example, that the long arms of the negro are closer to those of the great apes than the shorter arms of the whites, or that the apes have broad, flat noses as have also the negroes. But arguments of this kind depend for their validity on the particular anatomical characters that are chosen: if we take other characteristics we find that the whites approximate more closely to the apes than either of the other main human groups. Although it may be true that in the length of his arms the negro is closest to the apes, yet in the length of his legs he is furthest removed. Furthermore, the ape has practically no lips, and neither has the mongoloid: the white is intermediate in this respect, and the negroes are most "advanced". In hairiness of the face and body the white resembles the apes most closely, the negroids next and the mongoloids least. In hair texture the Caucasian is again most "primitive" while the frizzy hair of the negro is furthest removed from the apes. Again supra-orbital ridges—bony protuberances over the eyes—are found in the gorilla and chimpanzee and in many of the early human or sub-human remains. They, too, are far commoner among the Caucasians than among negroes. On balance, therefore, if one judges primitiveness by the amount of resemblance in physical characteristics between the races of mankind and the apes, the negroes are much the most superior race of the three.

A special form of the theory of racial superiority is to be found in the Nordic or Aryan[1] myth developed from the writings of de Gobineau[2] and Chamberlain.[3] De Gobineau wrote in defence of aristocracy at a time just after the revolutions of 1848. His writings were in no way nationalistic. He held that the rise of civilisations was due entirely to the predominance of an aristocracy—in whatever country—of Aryan (Nordic) type. The Alpines are of yellow descent and the Mediterraneans of black. These two groups form the proletariat, and if a group

[1] The term "Aryan" should, strictly speaking, refer to a group of languages and of languages alone. As Max Müller (32) p. 120 points out, to talk of an Aryan race is to make as big a mistake as to speak of a dolichocephalic dictionary or a brachycephalic grammar.

[2] De Gobineau (14). [3] Chamberlain (10).

destroys its aristocracy, the true descendants of the white race, that country is doomed.

A nationalistic slant to the theory was given by Chamberlain who glorified what he called the "Teutons". The definition of a Teuton was somewhat obscure, as Louis XIV, Michelangelo, Dante, Marco Polo, Paul and Jesus Christ were included. It depended on an unfaltering loyalty to a leader, a spirit of adventure and on martial virtues, and whether an individual manifested these characteristics or not had often to be determined by "intuition". "Whoever reveals himself German by his acts, whatever his genealogical tree, is a German." The Jews, on the other hand, revealed themselves by special ways of behaving. "One can very soon become a Jew . . . one needs only to have frequent intercourse with Jews, to read Jewish newspapers, and so on." Physical traits or genealogy do not distinguish them. One can see how much of this doctrine was incorporated into the teachings of Nazi Germany.

Now however little scientific basis there is for these theories or for their more recent developments (e.g. Madison Grant's[1] attempt to prove that the past and the future of America lies with the Nordic race), there is no doubt about the enormous attraction they have had and still possess. One of their most usual concomitants is the doctrine that the race must be kept pure, and the condemnation of miscegenation or racial mixture.

In spite of all the scientific evidence to show how great has been the intermingling between practically all the races of mankind, arguments are still abundantly produced to urge the necessity of keeping races—which are in fact of mongrel consistency—"pure". Let us take as an example of this the presentation of the problem by Davenport and Steggerda.[2] The main disadvantage which is alleged is that race crossing leads to "disharmony" in the offspring. If, for example, you mix a group with long jaws and teeth with another whose teeth and jaws are small the result may be a disproportionate effect in the offspring. Davenport, in fact, in an earlier book put forward the view that the prevalence of tooth decay in the United States (and presumably also in Europe) may be due to this cause. Similarly, if you mix a negro who has relatively long arms and legs with a white who has relatively short limbs the result may be an individual with short arms and long legs. Apart from looking odd such an individual would have to stoop more to pick things up

[1] Grant (17) and (18). [2] Davenport and Steggerda (13).

from the ground.[1] Davenport does not, however, consider the point that an individual with the opposite combination would have an advantage in this respect, such as it is.

It is true that in crosses between two very different species there is evidence that disharmonious combinations more often occur than harmonious. Jennings[2] gives an example in crosses between two species of fish, *Fundulus majalis* and *Fundulus heteroclitus*. The former is the larger and has larger eggs. In these eggs the circulation after development has progressed is more rapid than in the smaller heteroclitus eggs. Thus when the young fish are ready to swim about the yolk has been absorbed and does not interfere with the swimming. Now when these large majalis eggs are fertilised by heteroclitus the young hybrid develops up to a certain stage. But in the hybrid the blood circulation is slower than in the pure majalis. The result is that the yolk is only slowly absorbed, and when the young fish is ready to swim there is still a mass of unabsorbed yolk. This weighs it down and it sinks to the bottom and dies.

Again in crosses between St. Bernard dogs and Great Danes the hybrids develop but become partially paralysed when they are about three months old. In other crosses between different types of dog new combinations often occur. Some of these are beneficial, but when the stocks are very diverse the result is more usually disadvantageous, e.g. in a cross between a dachshund and a St. Bernard some of the offspring have the very short legs of the dachshund combined with the large body of the St. Bernard so that the body drags along the ground.

On the other hand the most usual result of crosses between different species, when those species are not too diverse, is that the offspring in the first generation show characteristics which are more or less intermediate in type between the parents, e.g. in crosses between very heavy and very light species the offspring usually turn out to be somewhere between the two parents in weight, but with all their parts in the usual proporations. Physical organs are not usually hereditarily transmitted as units.

Furthermore, as against the disadvantages of crossing it has been pointed out that a hybrid is sometimes superior to its parents in that it possesses what has been called "hybrid vigour" (*heterosis*). The offspring are often stronger and more vigorous than either parent.[3] It occurs in many varieties of cultivated

[1] Davenport and Steggerda (13) p. 471. [2] Jennings (22) pp. 261–3.
[3] Davenport and Steggerda found no evidence of heterosis in their investigation.

plants and domesticated animals, and it is found in crosses between wild forms as well.

Finally, it may be worth while to mention that the opponents of racial mixture are often to be found among the ranks of those who are most strongly opposed to inbreeding, holding that it leads to an increase in mental defect and other undesirable qualities. How the two apparently contradictory beliefs can be held at the same time is not made very clear. In fact inbreeding does not of itself lead to the emergence of undesirable qualities, as we saw in Chapter III in the pedigree of the Darwin family, though if an undesirable characteristic is present or occurs by mutation in some member of the group it will be more difficult to eliminate it from an inbreeding group than from one which mixes with others.

Mental Characteristics

So far we have been concerned with this question from the physiological side. In recent times, however, the development of tests of sensory abilities, of intelligence and of other characteristics has led believers in a theory of racial superiority to look for support for their beliefs from these tests.

So far as reaction times are concerned there have been a number of investigations which show[1] that the simpler peoples have a quicker reaction time than whites, and a far greater power of tactile discrimination. Furthermore, many of the reports about their powers of perception put Sherlock Holmes in the shade. Damoglou,[2] for instance, reports that some negro camel drivers can tell by a footprint whether the person who has passed is old or young, male or female, a young woman or a mother, and whether carrying a load or not.[3] Again Ranke[4] reports that certain natives can distinguish the sex of a deer from considerable distances by noticing the way it walks.

Such observations as these which one might have thought could be taken to indicate the superiority of the simpler peoples over the whites have in fact usually been interpreted to prove their inferiority, it being held, in perhaps a somewhat circular form of argument, that such abilities are more "primitive" and that therefore a primitive group would be expected to do better on them than a more "advanced" group. On the other hand, and

[1] See, e.g. Myers (33) p. 222. [2] Damoglou (12).

[3] It should be remarked, however, that checking the accuracy of such observations often presents difficulties.

[4] Ranke (37).

quite irrationally, when primitive groups have been found to have a lower susceptibility to painful stimuli,[1] or a lower sensitivity to auditory stimuli[2] this has often been interpreted as indicating that the whites are the superior group.

The development of standardised tests of intelligence gave rise to the hope in the minds of those people who were searching for means of measuring racial superiority or inferiority that here at last was just the thing they had been looking for. Psychologists who have devised intelligence tests have always had a very strong and sometimes irresistible tendency to claim that their tests are measuring innate intelligence. Intelligence is first defined in such a way as to indicate that it is a wholly inherited ability: tests are then devised which are given the name of intelligence tests, and the conclusion is left to be inferred that the intelligence tests are measuring this innate thing called intelligence.[3] The motivation of psychologists to indulge in this type of wishful thinking is undoubtedly very strong. It would be of enormous practical importance if they were in fact able to devise a direct and accurate measure of inherited capacity.[4] But there are good grounds, as we shall see, for the belief that they are over-optimistic. But let us leave this question on one side for a moment.

The results of intelligence tests applied to different races at first served to confirm the belief in their own superiority of the white—and often "Nordic"—psychologists who had devised the tests. During the war of 1914–18 tests were given to the recruits to the American forces.[5] Their results showed a very marked difference between the scores of whites and negroes, with only a 12 per cent. overlap between the groups. Furthermore, within the white group, those recruits whose origin had been Northern or Western Europe did far better than those coming from Italy or Poland—the two countries at the bottom of the list. Other investigations have shown that whereas the Chinese and Japanese

[1] McDougall (30) p. 195. [2] Bruner (9).

[3] For a further discussion of these points see my *Psychology and the Social Pattern*, pp. 61–77.

[4] This is not to say that intelligence tests as they exist at present are of no practical importance. On the contrary they are extremely valuable measures of the relative ability of different individuals when those individuals are drawn from the same race and social class and have been exposed to identical or very similar environmental backgrounds. That is to say, they may give us information about the innate capacities of individuals within a particular group; they will not, however, give us any direct information about the relative innate abilities of different groups. Their usefulness would be enormously enhanced if they could.

[5] Yerkes (41).

scored very little lower than the whites, the American Indians and the Mexicans gave scores which were as low as or lower than the negroes. During the last ten or fifteen years, however, the unquestioning belief in the accuracy of intelligence tests as measures of innate ability has been seriously challenged.

So far in this chapter we have been concerned with the examination of differences in characteristics between different groups of people without concern for the possible influence of environmental factors. In the case of intelligence tests it will be seen that cultural and environmental factors are in fact of very considerable importance. But before we consider this in detail, let us look back for a moment to see if environment has been shown to have any effect on the other criteria we have considered.

Environmental Influences

The differing effect of exposure to the sun's rays on the skin is well known to most members of the white race. Some freckle, others turn red, while others become gradually browner. The effect of the sun's rays on hair colour is less noticeably different, and its effect on eye colour is not apparent at all. So far as stature is concerned, a number of studies have shown that this may be influenced within limits by diet, by illness and by other environmental conditions. But one of the most startling and perhaps unexpected effects of environmental forces on a physical character has been shown to occur with cephalic indices. Gates[1] quotes a number of experiments in which the head form of babies has been modified by laying them on their back or on their side, e.g. of 555 babies born in Stuttgart in the years 1905–10 half were laid on their back and half on their side. Thirteen days later 84 per cent. of those laid on their back were more round-headed and 63 per cent. of those laid on their side were more long-headed than they were at birth. A number of them retained the determined form when they were investigated twenty years later. In another experiment a pair of monozygotic twins were both laid on their back, *A* on a soft cushion, *B* on a hard one, with the results shown in Table 8.

A still more striking indication of the effect of environmental influences on cephalic indices is to be found in the work of Boas,[2] subsequently confirmed in the main by Guthe,[3] Hirsch[4] and

[1] Gates (15) pp. 323–4. [2] Boas (6). [3] Guthe (19).
[4] Hirsch (20).

Spier.[1] Anastasi[2] reports that in Boas' investigation American-born and foreign-born boys were compared within the Eastern European Hebrew and the Sicilian groups living in New York

Age	*A's cephalic index*	*B's cephalic index*
1 day	83.5	87.1
18½ months . . .	86.2	78.4
19 years . . .	87.1	82.3

TABLE 8. Effect of lying position on head shape of a monozygotic pair of twins. (From Gates (15) p. 323.)

City. The Eastern European Hebrew is characteristically brachycephalic: the Sicilian group is characteristically dolichocephalic. But the Jewish boys born after their mothers had lived in the new environment for a number of years were more long-headed than the foreign-born Jewish boys, and the American-born Sicilian boys were more round-headed than the foreign-born—both groups, therefore, converging towards the American norm. In the second place, those born after a relatively long period of American residence of their mothers showed a greater change than did those born after a shorter period of residence. This is shown in Table 9.

	N	*Average age*	*Average cephalic index*
Foreign-born Sicilian boys .	241	9.6	79.5
American-born Sicilian boys:			
Born less than 10 years after arrival of mother . .	375	10.0	80.0
Born 10 or more years after arrival of mother . .	127	9.5	81.8
Foreign-born Hebrew boys .	179	9.1	84.6
American-born Hebrew boys:			
Born less than 10 years after arrival of mother . .	257	9.2	82.4
Born 10 or more years after arrival of mother . .	290	9.2	82.3

TABLE 9. Change in Cephalic Index of Two Immigrant Groups. (From Anastasi (2) p. 460.) (Data from Boas (6) p. 10.)

[1] Spier (40). [2] Anastasi (2) p. 459.

Furthermore, it was shown that these differences were not due to selective immigration, for a comparison of foreign-born persons who had immigrated at different periods showed no significant difference in cephalic indices, and, even more remarkable, the measurement of American-born and foreign-born children of the same parents showed differences in the expected direction.[1]

So far as the remarkable powers of observation of some of the simpler peoples is concerned it can be shown that many at least of them are due to training.

The Testing of Intelligence

Now let us turn to the testing of intelligence. Alexander[2] made a detailed analysis of the intelligence test scores obtained by the recruits into the American army who came from different states of the Union. He found that there was a high degree of agreement (correlation +0.72) between the relative educational efficiency of the states—judged by per capita expenditure on education, teachers' salaries, proportion of children attending high school and so on—and the intelligence test scores obtained by the recruits from those states. That is to say, ability to do well in the intelligence test depended partly, at any rate, on the efficiency of the schooling which a person had undergone.

Again, the negroes in the northern states have been shown to give a consistently higher score than those in the southern, whereas the whites in the South show a consistently lower score than whites in the North. Some writers have thought that selective migration might account for this difference—the superior negroes migrating to the North and the more inferior whites gravitating to the South. However, Klineberg[3] has conclusively shown that it is environmental forces rather than selective migration which are responsible. The school records of over 500 negroes who had left the South for the North were examined, and these showed that although there were differences between town and town, the average of the scores was almost exactly the same as that of the whole negro school population in the three towns studied. Furthermore, it was shown that the longer the period of residence in the North the greater—within limits—was the effect of the environment on raising the intelligence test

[1] For a somewhat critical account of Boas' results see Morant and Samson (44).
[2] Alexander (1). [3] Klineberg (24).

scores (whereas if selective migration were a factor, length of residence ought to have had no appreciable effect on the scores). A large number of Southern-born negro children aged 10 or 12 residing in the Harlem district of New York City were first equated for sex, type of school and socio-economic status, and they were then given a series of tests. The Stanford-Binet and National Intelligence Test results are presented in Table 10.

National Intelligence Test			*Stanford-Binet*		
Length of Residence	*N*	*Average score*	*Length of residence*	*N*	*Average I.Q.*
1–2 yrs.	150	72	Less than 1 yr.	42	81.4
3–4 yrs.	125	76	1–2 yrs.	40	84.2
5–6 yrs.	136	84	2–3 yrs.	40	84.5
7–8 yrs.	112	90	3–4 yrs.	46	88.5
Over 8 yrs.	157	94	Over 4 yrs.	47	87.4
Northern-born	1017	92	New-York-born	99	87.3

TABLE 10. Effect of length of residence in the northern States on the intelligence test score of Southern-born negro children. (From Klineberg (25) pp. 186-7.)

There is no doubt that the environment greatly affects the test score, particularly within the first few years.

Finally, the possibility was examined that these results might be accounted for by a deterioration in the quality of the later migrants. This was investigated by testing in two successive years groups of boys who had lived in New York City the same length of time. In this investigation it was found that the later migrants were *superior*, not inferior. Klineberg's work provided confirmation of previous work by Long[1] and McAlpin[2] which had suggested similar results.

A previous investigation had been undertaken by Klineberg[3] in order to study the problem of the relative levels of intelligence of Nordics, Alpines and Mediterraneans. It was mentioned on p. 81 that the American Army test showed that recruits from Northern or Western Europe scored more highly than those from Southern or Eastern Europe. Subsequently Brigham[4] worked up these results to show that the Nordic group was superior to the Alpine which was superior to the Mediterranean. His

[1] Long (28). [2] McAlpin (29). [3] Klineberg (23).
[4] Brigham (7).

method of evaluating the criteria for racial characteristics was, however, fallacious, as Brigham himself has subsequently admitted.[1] Klineberg studied 1,000 ten to twelve year old boys in rural areas of France, Germany and Italy. Only those children who had themselves as well as their parents been born in a particular area were chosen, and every child in addition had to possess the physical characteristics of eye colour, hair colour and cephalic index which are supposed to be typical of the appropriate race. The different groups were also equated as far as possible for economic, occupational and social level. The results are shown in Table 11.

Group	*Province*	*No. of villages covered*	*Performance scale scores*		
			Mean	*Median*	*Range*
German Nordic .	Hanover	17	198.2	197.6	69–289
French Mediterranean	Eastern Pyrenees	12	197.4	204.4	71–271
German Alpine .	Baden	10	193.6	199.0	80–211
Italian Alpine . .	Piedmont	10	188.8	186.3	69–306
French Alpine . .	Auvergne and Velay	19	180.2	185.3	72–296
French Nordic . .	Flanders	13	178.8	183.3	63–314
Italian Mediterranean	Sicily	9	173.0	172.7	69–308

TABLE 11. Intelligence test scores of Nordic, Alpine and Mediterranean children in Europe. (From Klineberg (25) p. 193.)

It can be seen that there were very considerable variations between different samples of the same "racial" group. The German Nordic group obtained the highest mean score, but the French Nordic was next to the lowest: the Italian Mediterranean group got the lowest score, but the French Mediterranean was among the best (getting in fact a higher median score than any other group). Again, there were very marked differences between the samples of the same race. Whereas in Germany the Nordic sample was superior to the Alpine, in France the Nordic was inferior to the Alpine and the Mediterranean was better than either. But in Italy the Alpine was superior to the Mediterranean. Finally, when all the racial groups, irrespective of nationality, were grouped together it was found that there were no significant differences between the scores of the Nordic, Alpine and Mediterranean groups.

[1] Brigham (8).

A study of the intelligence test scores of urban and rural groups also provides information which is of importance in the study of racial groups. We have already discussed this problem in Chapter III. It will be remembered that urban groups often score more highly than rural groups. To account for this the principle of selective migration is often cited. The more able individuals are expected on this hypothesis to seek their fortunes in the city. On *a priori* grounds, however, it might be just as reasonable to assume that the inferior and therefore presumably unsuccessful would migrate and become wage earners in the city, rather than those who were successful and had property and a standing in their own communities. Klineberg[1] has shown, however, that the principle of selective migration is not the explanation of the difference. He gave tests to 12-year-old negro boys, most of whom had been born in the country, who had migrated to three large cities (New Orleans, Nashville and Atlanta[2]). He also included a city-born group for comparison. He found, as is indicated in Table 12, that a better score on the test was obtained with increasing residence in the city.

Years of residence	*N*	*Average score*
One	39	38.3
Two	25	43.2
Three	36	44.7
Four	47	62.5
Five	52	56.2
Six	53	62.2
Seven or more	165	68.7
City-born	359	74.6

TABLE 12. Effect of length of residence in the City on the intelligence test score of country children. (From Klineberg (25) p. 197.)

Klineberg[3] also reports the results of an investigation on white children who had migrated from rural areas to cities in the North. He found that the migrants showed no superiority over those who stayed behind.

As we saw in Chapter III there is evidence that when a rural child moves to the city he has a good chance of improving his score on an intelligence test merely as the result of his changed

[1] Klineberg (24).

[2] The migration here was from one part of the South to another and not from South to North as it was in the investigation which has already been discussed on p. 84.

[3] Klineberg (25) p. 197.

environmental conditions which make it easier for him to devise the answers to the tests. If this is the case when the comparison is made between urban and rural children, how much more must it be the case when intelligence tests are used to measure the mental differences between races? Even between rural and urban groups the differences are smaller when performance tests of intelligence rather than language tests are given, and with some (e.g. the *Mare and Foal* test) the rural children show a definite superiority over the urban children. The language factor is of still more importance in the comparison between different races. Although it is quite obvious that verbal tests cannot be used when a group does not speak the language at all, yet it is equally important not to use verbal tests for comparative purposes when there is relative unfamiliarity with a language in one group compared with another. The question of literacy is not an absolute one, and there are all degrees of relative literacy down to complete illiteracy. The language handicap is perhaps most marked when it is present in only a mild degree, for in such cases it may be apparent neither to the examiner nor to the person tested that it exists at all. This is the most probable reason for the consistently lower scores on verbal intelligence tests shown by bilingual groups compared with monoglot groups,[1] for until the two languages have been completely mastered there is likely to be confusion between the languages as well as a smaller knowledge of either one. When, however, performance tests are used, the difference between monoglots and bilingual groups is reduced to inconsiderable proportions, and even in some cases reversed.[2, 3]

Apart from the language factor there are a few final points which illustrate the difficulty of obtaining comparable scores from different groups. The intelligence test situation for the average English or American school child closely resembles his ordinary school work, and he is therefore spurred on to do his best and to do better than others. But among many American Indian tribes such attempts to excel play no part in their traditional pattern of culture. They will not exert themselves

[1] See, eg. Saer (38), Smith (39), Pintner and Keller (35) and Mead (31).
[2] See, e.g. Jamieson and Sandiford (21) and Pintner and Keller (35).
[3] The widespread use of "progressive matrices" in testing recruits to the Army and Navy during the recent war may provide interesting further material on this point. The scale is non-linguistic and depends on the appreciation of perceptual relations (see Blackburn (5) pp. 66–8, Penrose and Raven (34)). It would be unwise, however to assume that even this scale measures the absolute amount of innate ability which a person may possess, for the effect of the environment is as likely to influence the appreciation of perceptual relations as the ability to perform any other abstractions.

therefore, to the same extent in the intelligence test situation.

Even when the intellectual operation required in the test situation is very similar to that which a person uses in the ordinary situations which he encounters in his everyday life, the task of abstracting the underlying principle and applying it in an entirely different context, namely that of the test situation, is, as Nadel[1] points out, one which varies in difficulty in different groups. In our society, for example, children are accustomed to play with coloured bricks and geometrical forms, so the material with which they are presented in performance tests contains few unfamiliar features. But among primitive peoples the abstractions required to carry out the test successfully have sometimes proved impossible, even though the test material has been carefully chosen so as to try to approximate to their cultural standards. One group of East African natives has been shown[2] to fail completely in the test of arranging coloured pegs in an alternating series. Yet it would be wrong to infer from this that they do not possess the necessary intelligence to perform a task of this type, for they plant trees according to the same principle in the course of their ordinary work with the utmost ease. It is true that one can argue that one of the essential features of intelligence is the ability correctly to apply general principles to relatively new situations, yet at the same time we should not forget that training in the technique of applying general principles to new situations may vary in thoroughness in different groups, and that in some it may be part of the cultural tradition, of education and of play. It is largely so in our own society. But in cultures where it is absent we should be wrong to expect the mental orientation towards tests and test situations to be the same as it is with us.

Again, among negro children the interest in intelligence tests is not so keen as it is among white children, even when the test is administered by a negro examiner. Where, as has sometimes been the case, the test is administered by a white examiner *rapport* is much more difficult to obtain. Special customs also tend to influence test scores. Porteus[3] found that it was very difficult to convince Australian aborigines that they were to do the tests on their own and without assistance, for tribal customs decreed that all problems should be discussed among all the members of the group until a unanimous decision had been reached. Again, among the Dakota Indians Klineberg[4] reports

[1] Nadel (43) p. 190.
[2] Bartlett (42) pp. 413–14.
[3] Porteus (36) p. 308.
[4] Klineberg (26) p. 155.

that it is bad form to answer a question if someone else who is present does not know the answer: and also that no answer shall be made unless the individual is absolutely sure about it. Thus compared with white children in England or America who often gain quite a few marks by lucky guesses when they think they know but are not absolutely sure of the answers, the Dakota Indians will be handicapped. One final example quoted by Anastasi[1] is worth recording. A young boy being tested in the Kentucky mountains was asked the standard question from the Stanford-Binet test, "If you went to a store and bought 6 cents worth of candy and gave the clerk 10c. what change would you receive?" The boy answered, "I never had 10c. and if I had I wouldn't spend it on candy, and anyway candy is what mother makes." The examiner then put the problem another way, "If you had taken 10 cows to pasture for your father and six of them strayed away, how many would you have left to drive home?" The boy answered, "We don't have ten cows, but if we did and I lost six I wouldn't dare go home." Finally the examiner made one last attempt, "If there were ten children in your school and six of them were out with the measles how many would there be in school?" The answer was, "None, because the rest would be afraid of catching it too."

Conclusion

Zoologists are faced with considerable difficulties when they try to find satisfactory criteria for distinguishing between different species: but these are small in comparison with the difficulties with which anthropologists have to contend when they try to find satisfactory criteria for distinguishing between different races. In the first place there are no groups of mankind that can be regarded as forming a pure line. Even if we allow ourselves to regard the few cases of socially or geographically isolated communities as forming a pure line, these are small in comparison with the great bulk of mankind. In the second place, the most obvious physical distinguishing marks, namely skin colour, eye form and colour, hair colour and form, stature and cephalic indices are phenotypic characters, and it is a dangerous and often unjustifiable procedure to regard these as indicators of true genetic differences. To classify races in terms of the relative proportions of people who belong to the four blood groups has the advantage of being a classification according to a genotypic

[1] Anastasi (2) p. 506.

character, but such a classification cuts across classifications based on other criteria, as, in fact, is also found when the phenotypic characters we have already mentioned are compared with one another. Furthermore, having decided which criteria we are going to use, we find a relatively small percentage of people who possess all the distinguishing marks, even if we use so few criteria that we get no more than two or three main races. A further difficulty is that environmental forces affect the different phenotypic characters to a greater or smaller degree. This is apparently even true of that sheet anchor of the physical anthropologist, the cephalic index. It is still more important in its effects on the measurement of intellectual characteristics—in particular in its effect on intelligence test scores. Here we can trace its influence on the scores of people brought up in places with different educational facilities and standards, on negroes who have migrated from the Southern to the Northern States of America, on urban and rural groups, on bi-lingual groups, and on groups with different systems of motivation, different customs and different interests. As a result of all these difficulties, the grouping together of people into "races" has to be viewed with considerable caution, and the attempt to prove the inferiority or superiority of some groups to others becomes meaningless unless very great care is taken in defining the exact characteristics which are being compared and the precise limitations of the comparative measurements.

REFERENCES

1. Alexander, H. B., "A Comparison of the Ranks of American States in Army-Alpha and in Socio-Economic Status", *Sch. and Soc.*, 1922, *16*, pp. 388-92.
2. Anastasi, A., *Differential Psychology.* New York: Macmillan, 1937, pp. 615.
3. Baur, E., Fischer, E., and Lenz, F., *Human Heredity.* London: Allen & Unwin, 1931, pp. 734.
4. Benedict, R., *Race and Racism.* London: Routledge, 1943, pp. 175.
5. Blackburn, J., *Psychology and the Social Pattern.* International Library of Sociology and Social Reconstruction, London: Kegan Paul, 1945, pp. 157.
6. Boas, F., *Abstract of the Report on Changes in Bodily Form of Descendants of Immigrants.* Washington: Government Printing Office, 1911, pp. 58.
7. Brigham, C. C., *A Study of American Intelligence.* Princeton, N.J.: University Press, 1923, pp. 210.
8. Brigham, C. C., "Intelligence Tests of Immigrant Groups", *Psychol. Rev.*, 1930, *37*, 158-65.
9. Bruner, F. G., "The Hearing of Primitive Peoples", *Arch. Psychol.* 1908, No. 11.
10. Chamberlain, H. S., *Die Grundlagen des neunzehnten Jahrhundents.* Munich: Bruckman, 1901, pp. 531.

11. DAHLBERG, G., *Race, Reason and Rubbish.* London: Allen & Unwin, 1942, pp. 240.
12. DAMOGLOU, C., "Hyperacuité olfactive et visualle et impassibilité a la douleur chez les nègres soudanais", *Rev. de l'Hypnot.*, 17, 1902.
13. DAVENPORT, C. B., AND STEGGERDA, M., *Race Crossing in Jamaica.* Washington: Carnegie Institute of Washington, 1929, pp. 516.
14. DE GOBINEAU, A. J., *Essai sur l'inégalité des races humaines.* Paris: Fermin-Didot, 1853, 2 Vols., pp. 492 and 512.
15. GATES, R. R., *Heredity in Man.* London: Constable, 1929, pp. 385.
16. GINSBERG, M., *Sociology.* Home University Library, London: Butterworth. 1937, pp. 255.
17. GRANT, M., *The Passing of the Great Race.* London: Bell, 1917, pp. 245.
18. GRANT, M., *The Conquest of a Continent.* New York: Scribner, 1934, pp. 393.
19. GUTHE, E., "Notes on the Cephalic Index of Russian Jews in Boston" *Amer. J. Phys. Anthrop.*, 1918, *1*, 213-23.
20. HIRSCH, N. D. M., "Cephallic Index of American-born Children of Three Foreign Groups", *Amer. J. Phys. Anthrop.*, 1927, *10*, 79-90.
21. JAMIESON, E., AND SANDIFORD, P., "The Mental Capacity of Southern Ontario Indians", *J. Educ. Psychol.*, 1928, *19*, 536-51.
22. JENNINGS, H. S., *Genetics.* London: Faber, 1935, pp. 351.
23. KLINEBERG, O., "A Study of Psychological Differences between 'racial' and 'national' groups in Europe", *Arch. Psychol.*, 1931, No. 132.
24. KLINEBERG, O., *Negro Intelligence and Selective Migration.* New York: Columbia Univ. Press, 1935, pp. 66.
25. KLINEBERG, O., *Race Differences.* New York: Harpers, 1935, pp. 367.
26. KLINEBERG, O., *Social Psychology.* New York: Holt, 1940, pp. 570.
27. KROEBER, A. L., *Anthropology.* New York: Harcourt Brace, 1923, pp. 523.
28. LONG, H. H., "The Intelligence of Coloured Elementary Pupils in Washington, D.C.", *J. Negro Educ.*, 1934, *3*, 205-22.
29. MCALPIN, A. S., "Changes in Intelligence Quotient of Negro Children", *J. Negro Educ.*, 1932, *1*, 44-8.
30. MCDOUGALL, W., *Report of the Cambridge Anthropological Expedition to the Torres Straits* (Ed. A. C. Haddon). Cambridge: University Press, 1901, Vol. 2.
31. MEAD, M., "Group Intelligence Tests and Linguistic Disability among Italian Children", *Sch. and Soc.*, 1927, *25*, 465-8.
32. MÜLLER, F. MAX, *Biography of Words and the Home of the Aryas.* London: Longmans, Green, 1888, pp. 278.
33. MYERS, C. S., *Report of the Cambridge Anthropological Expedition to the Torres Straits* (Ed. A. C. Haddon). Cambridge: University Press, 1901, Vol. 2.
34. PENROSE, L. S., AND RAVEN, J. C., "A New Series of Perceptual Tests: Preliminary Communication", *Brit. J. Psychol.* (*Med. Sect.*), 1936, *16*, 97-104.
35. PINTNER, R., AND KELLER, R., "Intelligence Tests for Foreign Children", *J. Educ. Psychol.*, 1922, *13*, 214-22.
36. PORTEUS, S. D., *The Psychology of a Primitive People.* New York: Longmans, Green, 1931, pp. 438.
37. RANKE, K. E., "Einige Beobachtungen über die Sehschärfe bei südamerikanischen Indianern", *Corr. Bl. Anthrop. Ges.*, *28*, 1897.
38. SAER, D. J., "The Effect of Bilingualism on Intelligence", *Brit. J. Psychol.*, 1923, *14*, 25-38.
39. SMITH, F., "Bilingualism and Mental Development", *Brit. J. Psychol.*, 1922, *13*, 271-82.
40. SPIER, L., "Growth of Japanese Children born in America and in Japan", *Wash. State Univ. Publ. in Anthrop.*, 1929, *3*, 1-30.

41. Yerkes, R. M., "Psychological Examining in the U.S. Army", *Memoirs of the Nat. Acad. Sci., Washington*, 1921, *15*, pp. 890.
42. Bartlett, F. C., "Psychological Methods and Anthropological Problems", *Africa*, 1937, *10*, 401-20.
43. Nadel, S. F., "The Application of Intelligence Tests in the Anthropological Field", Chap. 8 in *The Study of Society* (Ed. F. C. Bartlett and others). London: Kegan Paul, 1939, pp. 498.
44. Morant, G. M., and Samson, O., "An Examination of Investigations by Dr. Maurice Fishberg and Professor Franz Boas, Dealing with Measurements of Jews in New York", *Biometrika*, 1936, *28*, 1-33.

CHAPTER VI

NATIONAL DIFFERENCES

"An Englishman, a Frenchman and a Hebrew
Were sentenced to be hanged down Texas way.
When the Judge asked them which tree they'd like to hang on
The Frenchman chose the pear tree right away.

The Frenchman chose the pear tree and died happily,
The Englishman said any tree would do.
So on the apple tree they put his lights out,
When suddenly a voice came from the Jew.

'Hang me please on the gooseberry tree,
Won't you do this last favour for me?
It's my dying request,
It's the tree I love best,
Hang me please on the gooseberry tree'.

But the Judge said, 'You really must know
That the gooseberry tree is too low'.
'There's no hurry', said Mose,
'I will wait till it grows.
Hang me please on the gooseberry tree'."

R. S. S.

Nation and Race

Ginsberg[1] has defined a nation as "a body of people associated with, if not actually inhabiting, a certain territory, who have in common a stock of sentiments, thoughts and conative tendencies acquired and transmitted during the course of a common history and who have the will to be or to become politically independent, that is to exist as a separate state or to have some measure of autonomy, at least in cultural matters." From this it appears that the exact description of a nation is even more difficult than the exact description of a race. All the same it is possible to say that most of the criteria used for distinguishing between races concern hereditarily transmitted factors and physical differences, while most of the criteria used for distinguishing between nations concern environmental influences or psychological differences.

In popular thought the two terms race and nation are often confused and sometimes identified: Hertz[2] points out that people find it difficult to envisage the existence of a strong and cohesive bond between the members of a social group unless, as

[1] Ginsberg (6) p. 30. [2] Hertz (8) p. 52.

in the family, there is also a physical bond binding them together. The term nation is often used as if it implied some physical relationship, and in the term "the British family of nations" the idea is extended to cover a group of nations. It is possible that originally the word "family" was used in this context merely as a convenient term to describe the peculiar aggregation of nations in the British Commonwealth. Even so it has come to mean more than that in most people's minds, the connotation being extended to include a physical bond of unity, and this in spite of the enormous physical heterogenity within the different national groups.

Now although the terms *race* and *nation* are often confused and sometimes identified, it is desirable to draw a distinction between them. Although, as we have seen, an exact description of the two terms is a very difficult thing, yet they do refer to different aspects of human existence. Let us examine, then, some of the objective criteria which have sometimes been used to distinguish between nations.

Objective Determinants of Nations

It seems at first sight that some case might be made out for *language* as an essential determinant of a nation. Otherwise why should so much trouble have been taken in, say, Eire to revive their ancient and practically useless language? Again, so much passion would not have been spent over the teaching of German in the Italian Tyrol or in the Sudeten land unless the Austrians and Germans had thought that the "feeling of belongingness" to Austria and Germany could best be perpetuated in the minorities in these areas in this way. The national language, in fact, often comes to assume a symbolic importance, standing for national independence and honour and glory. It is generally assumed, as it was in the Peace Treaties of 1919, that people speaking the same language will want to form part of the same nation, and those cases where this is found not to be true are regarded as unexpected exceptions.

Yet though language is perhaps always an important factor in the determination of nationality it is not by itself a sufficient criterion. In spite of the existence of three languages in Switzerland the unity of the Swiss nation is not significantly impaired thereby. Nor is it in South Africa or Belgium where there are two official languages. And if one looks at the reverse of the picture the point is even clearer. The Americans and the English speak the same language, yet no one would claim (however much he might desire it) that they formed only one nation. Nor would he claim that because eighteen of the states in Central and Southern America speak Spanish, they formed only one nation.

The second objective criterion which needs to be considered is *religion*. It is probably true that if every nation had its own religion which was different from that of every other nation, the strength of the feeling of national unity and uniqueness would be increased. This is often implied in the utterances of nationalists, and indeed of many other people too when their country is engaged in a war with another whose religion is very similar to that of their own. When both sides are calling for the protection and succour of the same God it helps a little if it can be shown how one's own brand of religion differs from, and is a more accurate interpretation of, God's will than one's opponent's. Furthermore, there is a considerable body of opinion, at any rate so far as England is concerned, which holds that an Englishman who is a Mohammedan, a Jew, or even a Catholic cannot be a good and true Englishman.

In protestant countries the form the State religion takes is bound to be affected by the background of tradition, culture, mores and so on which already exist in that State. Each protestant nation, therefore, will be likely to have a religion which differs in many respects from that of every other. But this is hardly likely to be true of mainly Catholic countries such as Spain, Italy or Poland where the allegiance of the Church is to Rome, and where it is consequently more difficult for individual interpretations and practices to develop in different ways in the different countries. Certainly the mediæval attitude, as Hertz points out[1] implied that the unity of Christianity took precedence over national differences, and that the spiritual power of the Church was superior to the temporal power of the State. One's true fatherland was Heaven compared with which one's own country was of minor importance. Although this view is not so strongly held since the rise of nationalism, it still persists to complicate the loyalties of many Catholics to-day.

Thus we see that religion, like language, can be an important factor in the determination of nationality, though, also like language, it is not an essential condition.

Is the possession of territory an essential condition? Ginsberg in the definition already quoted does not lay it down as essential,

[8] Hertz (8) pp. 286–92.

though he maintains that there must be some association with a particular territory. Barker[1] maintains that the true nation has a home. If he had to invent a formula for making a nation he would say, "Take first a territory: add some form of organisation (or State) to hold its inhabitants together; let one language, if it was not there in the beginning, gradually prevail by its weight; let some community of belief and worship unite the spirits of men—and then from the crucible of time and the fermentation of the centuries a nation will emerge". A territory comes first for a nation, just as a home comes first for a man. Hertz[2] agrees with Barker and thinks it is essential. "We can imagine a nation without a state of its own," he writes, "but hardly one without a more or less contiguous area which it inhabits and in which it is rooted both physically and spiritually." The great difficulty about this view is clearly the case of the Jews which Hertz considers fully. He believes that although the Jews have once been a nation, and although their religion has often kept alive the hope that their nationality will revive, yet they should not be called a nation nowadays. Since the abolition of laws about territorial and social segregation the Jews have become increasingly assimilated and mixed with the populations in whose countries they are living, and they have in consequence tended to regard themselves as American, British, German or French by nationality and Jewish by religion. Many, in fact, have abandoned their religion in order to facilitate the process of assimilation. It is only, Hertz thinks, the rising tide of anti-Semitism which has led to the revival of Jewish territorial claims, and, in consequence, of Jewish national aspirations.

Yet these arguments do not, I think, dispose of the view that a nation can be formed of people who are not in possession of, nor desirous of possessing, a clearly defined territory. It may be true that many Jews regard themselves as American or British by nationality and Jewish by religion, but there are very many also who, without possessing any aspirations towards a Jewish national home, regard themselves as Jewish first and as American or British second. Furthermore, as Hertz[3] himself points out, in the former Austro-Hungarian Empire twelve nationalities were living together, yet the territory which each of these nationalities was desirous of possessing was not very clearly defined—as the draftsmen of the Peace Treaties of 1919 discovered and the subsequent historical events proved.

[1] Barker (1) p. 15. [2] Hertz (8) p. 146. [3] Hertz (8) p. 7.

I think it may be possible to go further than even Ginsberg has been willing to go and to visualise a nation as existing without reference to any clearly defined territory at all. I think it may be found that the strength of psychological bonds greatly exceeds that of physical bonds, and that none of the physical bonds, not even the possession of, or association with, a certain territory, is an essential feature of a nation.

Psychological Determinants of Nationality

Let us turn therefore to these psychological bonds. Hertz[1] points out that a person may legally belong to a nation which he fiercely hates or which does not regard him as a true national. It is, he adds, the social force of national consciousness rather than any legal concept which is the most important determinant of nationality. Feelings, emotions, interests and attitudes are stronger than any physical facts. There can be no nation without a sufficient measure of national consciousness, though it may sometimes be difficult to decide whether a particular group of people possesses this national consciousness in a sufficiently intense degree for it to be proper to regard them as a nation. Essentially, as Renan[2] pointed out long ago, it is the traditions and events of the past together with a desire to live together that forms the integrating bond in a nation. It is true that this bond is likely to be more easily formed when a group of people is geographically isolated or when it shares the same language or religion, but more important than any of these objective facts is the strength of the will to regard oneself as a member of a particular nation.

The strength of these psychological bonds may be seen when, for instance, a group of people is threatened, or believes itself to be threatened, with the destruction of its own ideas, traditions, customs and ways of life by the oppression of another group. In such circumstances the integrating bonds, the feeling of national self-consciousness, suddenly assume an overwhelming and (to many people) unexpected strength, as they did in the England of 1940. A common fear or a common distress is more likely to lead to a strengthening of these bonds than common victories and triumphs. When all is going well there is room for the discussion of a diversity of views, but in a terrifying emergency the only road to self-preservation is through the sinking of differences

[1] Hertz (8) p. 7. [2] Renan (13) p. 26.

and the discovery of the highest common factor of agreement.

It has already been mentioned that it is sometimes difficult to decide whether a particular group of people possesses national consciousness in a sufficient degree for it to be properly considered as a nation. But in some cases we can say that it is certainly not strong enough. As an example of this we may take the case of the Macedonians before the first world war.[1] Six nations claimed them—the Greeks who dominated the church and the schools and who said the civilisation of Macedonia was Greek: the Bulgarians on the grounds of language and sympathy: the Serbians on the grounds of language and folk-lore: the Albanians on the ground of race; the Rumanians on the grounds of language and culture: and the Turks on the ground of government. It is stated that a whole village would change over from one side to another frequently in order to gain economic advantages or sometimes to avoid being murdered. Thus none of the objective criteria were sufficient to decide the issue: it could only be settled by the existence of a spirit of national self-consciousness among the Macedonians themselves, and as this did not exist in the vast majority who tended to regard their village as their nation and not to see beyond it, the problem remained unsolved.

If the existence of a spirit of national self-consciousness is the essential link which binds a nation together, is there any way in which this feeling can be measured? Before the rise of democracy it could only express itself through the actions of the ruling and privileged classes in each state. The common people for the most part either did as they were bid, like the villagers in Macedonia, or tried to ape the manners and customs of the upper classes in so far as they were able. But the arrival of democracy, together with the use of instruments like the referendum and the plebiscite for the purpose of expressing the will of the people, opened new possibilities of measuring how people felt and thought and how they wanted to behave.

Nevertheless, how far such techniques can really be taken to demonstrate the wishes and aspirations, thoughts and feelings of the mass of the people is very debatable. It partly depends on the strength of the majority vote. A plebiscite in which the majority vote represents 51 per cent. of the total with the minority vote standing at 49 per cent. tells nothing very definite about the wishes of the community as a whole. This point has been fully appreciated by countries governed by dictators who have

[1] Cited by Hertz (8) pp. 241–2.

taken means to ensure that the majority is well over 90 per cent. of the whole. This is achieved by the illegitimate use of all the instruments of propaganda which the rise of democracy has brought with it—popular newspapers, radio, the cinema, popular catch-phrases, and popular education—together with the terrorising influences of military parades, secret police and an open ballot. Without these last three items and with the same powers to make use of all the other instruments of propaganda given equally, and for a considerable period of time before the plebiscite is taken, to all parties, it might only be on rare occasions that an overwhelming majority of the kind obtained by the plebiscites conducted under dictatorships would be encountered. There would almost certainly be no overwhelming majority in these circumstances in the case of the Macedonians: would it also be true of the English or the French or the Germans or the Americans or of many of the other nations of the world when the issue on which the vote was taken concerned their nationality? In such cases there would be little hesitation and practically no disagreement among the members of those nations about which nation they belonged to. A plebiscite on the question conducted in the most complete secrecy and without any terrorisation would give a result comparing favourably with the best that Hitler ever achieved. Very many years of propaganda—it might even be true to say very many years of one-sided propaganda—would be necessary before any substantial difference was recorded in the voting. In these cases, therefore, the method would be meaningful: it would show the existence of a very strong feeling of national self-consciousness. But in other cases the method becomes almost meaningless, and is bound to remain so until we know how large the majority vote must be before we can say that the feeling of national self-conciousness represented by this majority is strong enough for true nationhood to exist.

National Character

It it can be assumed that a group possesses a sufficiently strong degree of national self-consciousness for the group to be properly considered as a nation, is it justifiable to proceed a step further and endow this nation with national character? Here at once we encounter very considerable difficulties. How far such characteristics as may be believed to be representative of a particular nation really will be will depend on a number of factors.

The first three, and perhaps most fundamental of these

factors which may affect the way in which a group of people lives and the characteristics they develop depend on geography, economics and the cultural pattern. Man may alter the circumstances of his geographical environment, of his economic environment, or he may select some modes of behaviour rather than others and weave them together into a cultural pattern. But it is also true to say that man himself may be altered by the different opportunities with which the geographical or economic environment provide him, and he is certain to be affected by the traditions and the customs of his particular cultural pattern.

(a) *Geography*

Man's physical environment is not, as Barker[1] points out, a constant quantity. "His environment changes—changes because he modifies its nature by his own action (when he clears the forest or drains the marsh); changes, again, because he alters the direction of his attention—allowing himself to be influenced now by this factor, and now by that—according to his needs, or his policies, or his inventions, at any given moment of time."

Furthermore, "the physical factors of his environment exert their influence not by affecting his body, but by entering into his mind and becoming motives of action; and the ideas which he forms may modify, or even contravene, the influence which we might expect any given factor to exercise. He may refuse an obvious food or a 'natural' way of life from some notion of taboo, or some obstinate adherence to a previous tradition; and far from being the servant of environment, he either may pit himself against its suggestions, or, at the most, condescend to an independent co-operation."[2] Once again we may see that it is not the innate agencies alone, nor the environmental agencies alone, which will determine the characteristics which a group of people may show. Both are necessary, and what the one may do in the case of one group may be done equally well by the other in the case of another group.

It is true that some writers in both ancient and more modern times have elaborated a theory of geographical determinism.[3] According to this view the physical environment in which men live is a constant factor and a stable force, and it is this which leads to the fact—according to these writers—that history repeats itself over and over again. Island peoples show fundamental

[1] Barker (1) p. 48. [2] Barker (1) pp. 48–9.
[3] For a full discussion of this theory see Semple (14).

points of agreement in their characteristics which differ from those shown by peoples who live in deserts or steppes or river valleys or mountains. Similarly, too, the climate in which people have to live is a determining factor in their way of life, and consequently in their character. Climate includes humidity, and the amount of sunlight, and it includes also winds and storms.

But it is possible by selecting one's instances with sufficient care to prove almost any theory: it is equally possible to prove the opposite. Thus Aristotle believed that the temperature of Greece was the best for producing an ideal people: Ellsworth Huntington[1] on the other hand produces arguments to show that the climate of England is the best in the world.

If the matter is looked at more broadly, as for example by Forde,[2] it can be seen that the occupants of regions which are alike in their geographical conditions often differ very markedly in their cultural achievements, while those who live in regions having very different geographical environments are often very similar in their economies and culture.

It is better, therefore, to approach the question of the influence of the physical environment on the characteristics of the people inhabiting that area from the point of view of *possibilities* rather than from the point of view of determinism. Thus an island offers different possibilities to its inhabitants from a continental country, for it offers possibilities of development by sea as well as by land, and if an island people turns its back on the sea with its hazards and adventures it is very likely to develop in a very different way, and so to show very different characteristics, from one whose main preoccupation is in ship building, exploration and overseas trading.

In the course of history a group of people may choose one set of possibilities at one period and quite a different set at another, and to the extent that it does so, it will affect its predominant ways of behaving. The influence of the shape of a country, its climate, and its natural resources must not of course be neglected, but the important thing when one is considering their effect on the behaviour of the people who live amongst them is to try to see how those people have used the possibilities they have been given, how they have modified them (and so perhaps modified the range of their possibilities) and how they have counteracted them. For some influences may be counter-

[1] Ellsworth Huntington (9) pp. 220–26. [2] Forde (5).

acted. Thus great heat or intense cold may be modified by air conditioning: natural resources of water power in inaccessible areas may be harnessed, have their energy turned into electricity, and transmitted to centres of population where they may be more usefully drawn upon: and even the damp climate of Lancashire which has been regarded as of so much importance in the spinning of cotton, may be reproduced and improved upon by artificial means closer to the source of the raw material.

(b) *Economics*

From the geographical possibilities we turn to the economic, of which population and occupation are two of the most important. A densely populated area will tend to produce different sets of characteristics from a sparsely populated area: a predominantly agricultural people will tend to develop differently from a predominantly industrial people. To some extent these two factors are interdependent. A densely populated area must turn to industrial occupations and exchange its products for food, for there is not enough room for them all to gain their livelihood from the land. And as the density of population in a particular area changes, so will the predominant occupations within that area change also—and so will the characteristics which that group of people will show.

In addition to the quantity of the population we need to consider its quality. A group of poor mental and physical stock shows far less initiative, far fewer ideas, and far smaller ability to select from the possibilities with which it is surrounded, and a much reduced ability to cope with the changes in its environment than a group of better stock. We need to consider also the degree of urbanisation in a country, for this will affect the usual way of life of its inhabitants. A country which is 80 per cent. urbanised, like England, will tend to direct its way of life much more exclusively to the town dweller and to neglect the interests of the minority of country folk, than will a country like France which is still predominantly agricultural.

Finally, it is important to consider the relative proportion of the sexes, and still more important the age distribution. A country in which there is a majority of people under 40 will be likely to show different characteristics from one in which the majority are over 40, and the age distribution is a factor that is constantly changing. When the population of a country is growing fast there will naturally tend to be a predominance of youth over age: in a country whose population is stable or declining the needs of the older people become of growing importance, and not only are the needs altered, but the outlook of the inhabitants is likely to be different too when the population is older—they will have different interests and different requirements.

(c) *The Cultural Pattern*

The third fundamental factor which needs to be considered when one is investigating the question of national characteristics is the pattern of culture of the particular group being studied. This approach has been made by Mead[1] in extending the theory which was given wide currency by Benedict[2] in order to explain and interpret the American character.

Ruth Benedict emphasizes the importance of customs in holding society together and in determining the way in which the individuals within that society will behave. Her view is that the binding force of customs and of the pattern of a culture is stronger than that of biological heredity or than that of any other single factor. She points out that the diversity of cultural patterns is almost limitless. What in effect happens, she argues, is that societies make selections out of all the possible ways of living and behaving and weave these selections into a pattern which henceforward largely determines the actions of their individual members. Thus, "One culture hardly recognises money values, another has made them fundamental in every field of behaviour. . . . One builds an enormous cultural superstructure upon adolescence, one upon death, and one upon adult life."[3] Another example is warfare. The Aztecs used warfare simply as a means of getting captives for religious sacrifice; they did not fight to kill, and consequently when the Spaniards appeared on the scene and fought to kill, the Aztecs retired in disorder. In other groups all idea of warfare is ununderstandable: Western methods of warfare are to them on much the same level as a disorderly house brawl is to us. Some groups cannot understand periods of peace alternating with periods of war, for a period of peace seems to imply that the enemy can be treated as a human being, an idea they cannot share.

Similarly with homicide. "We might suppose that in the matter of taking life all peoples would agree in condemnation. On the contrary, in a matter of homicide it may be held that one is blameless if diplomatic relations have been severed between

[1] Mead (12). [2] Benedict (2). [3] Benedict (2) p. 24.

neighbouring countries, or that one killed by custom his first two children, or that a husband has the right of life or death over his wife, or that it is the duty of the child to kill his parents before they are too old. It may be that those are killed who steal a fowl, or who cut their upper teeth first, or who are born on a Wednesday. Among some peoples a person suffers torments at having caused an accidental death; among others it is a matter of no consequence. Suicide also may be a light matter, the recourse of anyone who has suffered some slight rebuff, an act that occurs constantly in a tribe. It may be the highest and noblest act a wise man can perform. The very idea of it, on the other hand, may be a matter of incredulous mirth, and the act itself impossible to conceive as a human possibility. Or it may be a crime punishable by law, or regarded as a sin against the gods."[1]

In spite of the possible diversity of customs it must be remembered that each pattern of culture has an integrating thread running through it. Each pattern forms a unity of its own. A pattern of culture may be regarded as a Gestalt,[2] held together by its own internal stresses and tensions, and differentiated from the internally coherent pattern of another society. "Every little tribe awards special approval—some to this achievement, some to that—and the tribal institutions allow men to pursue the favoured goal. If this goal is deeds of daring the individual character fostered in that tribe is hardy and scornful of pain; if the goal is to raise a good crop in one's garden man is industrious and patient of routine. There is incredible variety in these goals, and most societies honour several different kinds of achievement, usually compatible one with another; whatever the goals are, and the social institutions which allow people to pursue them, human behaviour is modified in conformity with them.

The social institutions of any one tribe are not inevitably expressions of racial genius, for other tribes of the same race have radically different ones. The rugged individualism of American

[1] Benedict (2) pp. 45–6.

[2] The term "Gestalt" which has passed into the technical vocabulary of English-speaking psychologists, in spite of various attempts to translate it into the English "shape", or "configuration" represents the approach to psychology of a German school of psychologists of whom Wertheimer, Koffka, Köhler and (more recently) Lewin are exponents. The fundamental approach of these psychologists is that more valuable psychological material can be obtained by considering behaviour as it is actually experienced in all its complexity as the primary datum, and to start further investigation from this point, than to begin with elements and try to build up the complexities of behaviour from combinations of the elements. The whole is different from the sum of its parts: the whole or "Gestalt" determines the nature of the elements rather than the elements determining the nature of the whole. See Blackburn (16), Koffka (17) and Köhler (18).

Indian tribes of the western plains on the one hand, and, in great contrast, the mildness and sobriety of some Pueblo Indians on the other, are both responses to the different ways of arranging social life, and the different opportunities for the individual provided by the two cultures. Both groups are racially American Indians."[1]

These three factors are, then, of fundamental importance when one is considering the question of national character. All three must be taken into consideration, and it is necessary to try to avoid using any one of them alone as a single and complete explanation.

Apart from these three Ginsberg[2] has drawn attention to four others which will also affect the development of national character. One of these is the type of political organisation existing in the nation. We have already pointed out that characteristics which may emerge in a nation governed by a dictatorship with its concomitant system of spies, secret police and other techniques of terrorisation, cannot be taken to be representative of the whole nation. They will only be something imposed on the majority by the political domination of a minority. This is true in general, but the statement needs some slight qualification, for if a particular form of dictatorship were to last for thirty or forty years, then those characteristics which the dictatorship is trying to impress on the mass of the people might be assimilated by the rising generation to such an extent that at the end of that period the characteristics could be regarded as being representative of the nation.

So far as democracies are concerned there is likely to be less homogeneity in a nation whose form of government is of the federal type with wide legislative and administrative powers given to the several states of which the federal union is composed than there is in a nation whose form of government is more highly centralised.

In the second place the representative nature of national characteristics will depend on the form of class structure in the nation. We have already mentioned that in an undemocratic nation the characteristics which are likely to be held to be representative of the nation will be mainly those which are characteristic of the governing and privileged classes—the rest of the population for the most part doing as they are told or apeing the upper classes. Thus Ginsberg[3] remarks that most of the

[1] Benedict (3) p. 90. [2] Ginsberg (7) p. 185. [3] Ginsberg (7) p. 186.

characteristics attributed to the Poles appear to refer only to the Polish nobility, while of the bulk of the population we know little or nothing. The same thing is true to a lesser degree in many of the traits which are held to be characteristic of England. Many of the characteristics, in so far as they have been true at all have been based on a particular type of Englishman given a highly specialised and unrepresentative type of education at one of the dozen or so older public schools and one of the two older universities. How far such characteristics are national will depend on whether the rest of the population has attempted to copy the manners and customs of the privileged classes and on its success in doing so.

In the third place it is necessary to know the degree of cultural homogeneity which exists within a nation. In so far as this concerns the general level of education it will be bound up with the type of class structure in the nation. But it has other implications as well. One of these concerns language, which we have already considered as a determining factor in a nation. The point here, however, is that all languages carry with them a background of cultural heritage, so a nation which is unilingual is likely, other things being equal, to be more homogeneous and therefore to display characteristics more truly representative of the nation as a whole, than one which is multi-lingual. The second implication is religion. This again is a factor which we have already considered separately. But, like language, religion carries its background of cultural heritage, and, as with language, it can be said that, other things being equal, a nation which has only one religion will be more homogeneous, and will as a result be more likely to display a national character truly representative of all sections, than one which has many religions, and therefore many different cultural forces moulding the behaviour of its inhabitants.

In the fourth place Ginsberg maintains that the stage of maturity of a nation affects the display of national characteristics. To some extent this is true. Thus those nations which form a part of what we call Western civilisation are likely to show certain similarities in outlook, attitude and behaviour as a result of the interchange of ideas and influences which have developed, and which continue to develop, as Western civilisation proceeds on its career of materialistic and other forms of progress. And such similarities as these nations share as a result of these influences may be different from those betrayed by other nations forming a part of some other form of civilisation. Yet this is not to say, of course, that all Western nations display identical national characteristics: it is only one aspect of their behaviour which is affected. This, too, needs to be borne in mind when this line of argument is applied to "primitive" peoples. Thus Ginsberg says,[1] "The traits attributed by Tacitus to the Germanic tribes have often been regarded as specifically German; but according to modern ethnologists these traits are common to many primitive peoples." It is, however, necessary to add to this that although primitive peoples may show some similarities in outlook and behaviour as a result of the not very highly developed stage of maturity which they share in common, this does not imply that all primitive peoples betray characteristics which are identical in every respect. In fact, as we have seen above,[2] there is as much diversity in cultural pattern among primitive peoples as there is among the nations of Western civilisation.

In spite of these difficulties, and in spite of others which we shall consider in a moment it seems unreasonable completely to dismiss the concept of national character as an illusion. As Ginsberg[3] says, "Nations like other groups behave in distinctive ways and if their behaviour shows some unity and continuity we may perhaps speak of a group character without committing ourselves to any theory of a group mind or group personality." The notions of unity and continuity of behaviour imply some measure of stability in the government, class structure, culture and level of maturity of the particular nation whose characteristics are being studied. After a social revolution, or after a serious defeat in war, the old attitudes and habits and mores may be thrown overboard, and a period may ensue during which there are no clearly defined patterns of behaviour characteristic of the nation as a whole. Subsequently, when a new period of stability has lasted for some time, another national pattern may emerge which is fundamentally different from that which was characteristic of the old order. It would be unwise to assume that the national characteristics, whatever they may have been, of Germany in 1930 were the same as those of Germany in 1910 or in 1940, or that the national characteristics of Tsarist Russia were the same as those of the Soviet Union. The national pattern may change fundamentally as the result of the sudden shock of social revolution. It may, however, also change fundamentally, though more gradually and slowly, through the normal process of social evolution. The notions of unity and continuity are therefore

[1] Ginsberg (7) p. 185. [2] pp. 104–5. [3] Ginsberg (7) p. 188.

relative. A fundamentally different picture may be presented after a shorter or after a longer period of time, depending on whether the change has come about through social revolution or social evolution. Ginsberg mentions the notions of unity and continuity because they are included by Stout[1] in his definition of individual character. I am not myself sure, however, that the notions are fundamental to any definition of character. Individuals, like nations, can show fundamentally different patterns of character at different periods of time. Normally the process continues gradually as the result of a person's growing wealth of experiences, but his character may fundamentally change in the process so that the character he exhibits now may be entirely different from that which he exhibited twenty years ago or which he will exhibit twenty years ahead. Similarly, too, individuals, like nations, can entirely change their characteristics by a sudden process of "conversion". The nation employs the method of social revolution to achieve this sudden change: the individual usually undergoes some violent emotional shock or vivid experience after which he emerges as a person showing an entirely different character from that which he exhibited before. I think, therefore, that it is more meaningful to describe character (whether individual or national) in terms of the particular pattern of behaviour which the individual or the nation shows at a particular period of time rather than to stress the notions of unity and continuity, or to search for fundamental, underlying and perhaps innate character traits. As Ginsberg[2] says in the last sentence of his paper, "The national character is not something given once and for all, but something always in the making, moulding and being moulded by the circumstances in which nations find themselves."

I have argued elsewhere[3] that it is more useful to consider individual character in terms of the interests, sentiments and attitudes which a person builds up in the course of his existence in a particular community, rather than in terms of the underlying pattern of instincts and reflexes with which he is born. For one thing such an approach makes it easier to follow any changes in character which the person may display as the result of the impact of new experiences on him. It is true that his present reactions or his present character will be partly determined by his innate equipment of genes and chromosomes, reflexes and

[1] Stout (15) p. 653.
[2] Ginsberg (7) p. 204.
[3] Blackburn (4) pp. 124–31.

instincts, but this innate equipment remains the same whatever his experiences may be. So it is misleading rather than helpful to attempt to reduce individual character into these innate elements, unless one is trying to show that an individual's real character never alters. Such a position would be an unusual one to adopt, and even if it is adopted it makes it necessary to use other terms to describe that aggregation of traits in a person which lead him to behave in a relatively consistent way (even though it may be a consistent inconsistency) at a particular period of time, even though his behaviour at a previous or subsequent period of time may show quite a different picture. Now precisely the same arguments may be used when one turns to the question of national character. It is more useful practically to describe national character in terms of the predominating patterns of behaviour at a particular period of time rather than to search for unity and continuity or to look for fundamental underlying and perhaps innate character traits. My emphasis in this respect is somewhat different from that of Ginsberg who in some parts of his paper implies that it is necessary to search for the underlying similarity in the character traits of the members of a nation before one can properly understand that nation's character.

Ginsberg[1] rightly points out that the relationship between individuals and their institutions is one of reciprocal interaction —that although men make their instititutions yet institutions also make their men, and he considers it misleading to say with Madariaga[2] that it is not English education which explains the Englishman but the Englishman English education. This is true enough, though it may be important to point out also that the forces within the reciprocal interaction are continually changing direction and strength. Thus, particularly in conditions of political, social, cultural and economic stability, it is likely that the relative strength of the institution is far in excess of that of the individual, and in these circumstances it would be the Englishman who would be moulded by English education to a far greater extent than English education would be moulded by individual Englishmen. It might even be true that the weight of institutional influences is almost always in excess of that of individual forces, for the younger individuals with ideas for a radical change in the institutions do not usually possess the power to alter those institutions, and by the time they have attained a position of power they are usually a great deal older, less experi-

[1] Ginsberg (7) p. 188. [2] Madariaga (11) p. 13.

mental in their ideas and so far under the influence of the institution whose mores they have been forced to adopt while they had no power to change them that they do not wish to make any violent alterations in them now. It is only in conditions of violent social, economic or political upheaval of the kind already mentioned that there is likely to be a general abandonment of the institutional ways—the "bad old ways"—and a chance for the individual's power very greatly to exceed that of the institution's. Then it will be the task of the individuals to make new institut-tions—which will subsequently pass into a position of pre-eminence and mould the behaviour of the individuals again.

This however has taken us from the point we were considering which was Ginsberg's search for fundamental elements in national character. Thus Ginsberg[1] proceeds, "The situation is complicated by the fact that institutions may not reflect the character of all the members but perhaps only of powerful sections, and that, once formed, they tend by unconscious processes to select the type that suits them. In this way many qualities in a population may remain dormant or repressed until a change of circumstances brings them into play, while again the reaction against a type that has become unduly dominant may sometimes be an important cause of change. *It is thus possible for great and revolutionary changes to take place in the institutions of a nation without a parallel change in the underlying qualities in the mass of the nation.* It follows that at any one moment of time we cannot safely infer the character of a people from its institutions or public policy, and that for this purpose it is necessary to know the history of the institutions and the portions of the people that have been dominant in shaping them."[2]

Ginsberg argues[3] that a great deal of what foreign writers have to say about the English character may be inaccurate for it is based on their impressions of the Public school mentality, and the Public school tradition dates only from the middle of the nineteenth century, and few of the "Prep" schools which are an integral part of the system date further back than 1870. It is at this point that I believe Ginsberg's argument to be mistaken. In my view the relative recency of the Public school tradition is irrelevant to the question of the English character at the present time. For after all it *is* a relative recency: nobody who went to one of the older Public schools at the time the new tradition was being established is alive to-day. It is not as though the Public

[1] Ginsberg (7) p. 188. [2] My italics. [3] Ginsberg (7) pp. 188–9

school tradition was something that had sprung up within the last twenty years and had not therefore had time to influence the behaviour of the bulk of the population. If, for example, it is true that there is such an entity as the Public school mentality and code of behaviour, and if this institution has been developed by the most influential members of the English community, and if it carries a number of privileges with it, it would not be surprising to find many other non-Public school groups within the community assuming that the acquisition of an influential job depended in part on the display of this particular aggregation of characteristics, and in consequence attempting to divert their behaviour into the Public school pattern. And if they should be successful in doing so, then I think it would be fair to describe these characteristics as representative of the English national character at that particular period of time. In fact, of course, it is not so because the Public school code of behaviour was not universally nor successfully adopted by the rest of the population. The foreign writers may therefore be justifiably criticised for describing the characteristics of a part of the nation as though they were true of the whole, but it is unjustifiable to criticise them on the ground that they were basing their conclusions on too recent a tradition.

Nevertheless Ginsberg is perfectly right in emphasizing the importance of a study of the historical development of the institutions and culture of a nation, of its proverbs, folk-lore, wit and humour in order to appreciate properly how far the characteristics displayed by a pre-eminent group within that nation are really representative of the whole body of people. It is a line of study which awaits development. Ginsberg employs the historical method himself in his study of the character of the English and of the Germans. There is universal agreement among observers, both English and foreign, he says, that two outstanding characteristics of the English are empiricism and individualism. Empiricism is to be found in all spheres of English life, not merely in the contribution of English thinkers to empiricism as a philosophical theory. Thus in English law and English politics, in the history of the Church of England and in English international policy, there is a tendency to deal with particular problems as they arise rather than formulating general principles or indulging in long-range planning. In this respect the English can be sharply differentiated from the French who prefer general, clear-cut and logically consistent solutions to their problems.

The second characteristic of the English is their individualism, and this can also be traced through various spheres of life. It may be seen in the emphasis which is laid in law on the liberty of the subject, in the preference for decentralisation in government, in the strength of non-conformity and puritanism in religious matters. This individualism is closely bound up with the empirical approach and the capacity of the English for spontaneous organisation may rest on a combination of these two characteristics. Ginsberg argues that the history of Trade Unionism, of the Co-operative movement and of the Friendly Societies indicates that these characteristics are not peculiar to the governing and privileged classes in England.

The Germans, too, Ginsberg continues, are held by many writers to be highly individualistic. But the individualism of the Germans shows certain differences from that of the English. The German individualism is not combined, as the English form of individualism is, with the capacity for compromise and for achieving spontaneous organisation in order to solve immediate and practical problems. The organisation which the Germans achieve as a result of their individualism is an organisation based on relations of superiority and inferiority rather than on mutual co-operation.

Another characteristic of the Germans, to be seen in their economic life, in their industry, and in scientific work is the persistent searching for grandiose schemes. This leads them to combine profundity and comprehensiveness on the one hand with a measure of vagueness and obscurity on the other.

Much more can be said, and has been said by Ginsberg and other writers about the characteristics of the German and the English which emerge from a study of their institutions. But this indirect method of studying national characteristics, though very fruitful, is not the method which has been most usually employed. Far the more usual method is to compare different nations by endeavouring to measure their relative strength or weakness in the possession of different characteristics. This method is far less exhaustive and far more superficial than the study of national characteristics through the history and development of institutions. It also suffers from several other disadvantages. In the first place, such characteristics as are attributed to different nations by this method will be likely to be far more transitory in nature. When one studies national character by studying the institutions of a nation it is likely that something relative permanent will emerge (provided that the political, economic and social life of the nation being studied is not subjected to sudden and revolutionary changes). For, as Ginsberg[1] points out, when once certain institutions have been chosen by the individuals in a nation then it will not only be true to say what we said before, namely that the behaviour of everybody who comes in contact with those institutions will be moulded so as to conform to the pattern of characteristics which those institutions emphasize, but it will also be likely that those individuals whose innate characteristics are of the kind that are emphasized by the institutions will (because they will need less moulding than other individuals whose innate characteristics are different from or conflict with those emphasized by the institutions) tend to become the leaders in that nation and hence to reinforce the existing institutional pattern, whereas those whose innate characteristics conflict so strongly with those emphasized by the institutions that they remain in conflict in spite of the institutional pressure, may be expelled from the nation and become refugees, or they may be placed in concentration camps or in mental hospitals and so be unable to exert any pressure towards an alteration of the institutional pattern.

At all events when one studies national character by describing superficial traits there is no doubt that more differences appear at different periods of time than when one studies it through the historical method. Thus Hertz[2] points out that in former times the English had the reputation of being an unruly, revolutionary, fickle people while the French prided themselves on their loyalty to their kings and on the stability of their institutions. Again, Ginsberg mentions that although the characteristic of energy is now commonly attributed to the English by continental writers, yet in the eighteenth century it was stated that the greatest example of human indolence was to be found in England. Similarly, the cleanliness which is sometimes attributed by Englishmen to their Nordic ancestry is a relatively recent development. Again, the arrogance and pugnacity attributed to the Swedes even as late as 1730 would hardly be attributed to them to-day. Again, as Benedict[3] points out, history records very rapid changes in social behaviour—far too rapid to be accounted for on a genetic basis. "The England of Elizabeth and Shakespeare changed in one man's lifetime to the England of Cromwell." Japan, too, which on account of her premeditated

[1] Ginsberg (6) p. 36. [2] Hertz (8) p. 42. [3] Benedict (3) p. 81.

isolation up to 1868 had become much more stable racially than any nations in Europe, changed all her national characteristics within two generations. And the most dramatic change of all occurred within a generation or two among the Negroes who were taken to America as slaves. "Most of them had been transported from Nigerian kingdoms with prized cultural achievements. Their elaborate and ceremonious political organisation, the pomp of their courts, the activity of their far-flung economic life with its great market centres and tribute collected over great areas, their legal system with formal trial of the accused, with witnesses and with prosecutors—all these excite the admiration of any student. Belatedly we admire to-day the incisive folk-tales of Nigeria, their rhythmic dances, their wood carving that has excited the respect of modern European artists. But these are to-day collected in Africa. In America all this achievement was stripped from imported slaves as if they had never had part in it. Their patterns of political, economic and artistic behaviour were forgotten—even the languages they had spoken in Africa. . . . Conditions of slavery in America were so drastic that this loss is not to be wondered at. . . . It is no wonder that their owners remarked upon their lack of any cultural achievements; the mistake they made was to interpret the degradation of the slave trade as if it were an innate and all-time characteristic of the African Negro."[1]

Stereotypes

Generalisations which are made about the relative strength of different characteristics in different nations are also more likely to be influenced by personal bias and prejudice than are studies of the character revealed by the history of their institutions. It is usual to find that writers overestimate the extent to which their own nations possess desirable characteristics. Even British weather and British buns have been claimed by some writers to be the best in the world. Correspondingly, there is a tendency to underestimate the extent to which other nations possess desirable characteristics and to give them more than their due of the undesirable. However, such generalisations as are made, although they are of no value as descriptions of innate behaviour patterns may yet be of interest, and perhaps even of importance, as reflections of the attitude of groups towards one another at a given period of time. Now what do these generalisations amount to? In most cases they take the form of what are

[1] Benedict (3) pp. 84–5.

called "stereotyped attitudes". Thus it has been shown in the United States that the Germans are regarded as being predominately scientific, industrious and stolid: the Negroes as superstitious, lazy and happy-go-lucky: the Irish as pugnacious, quick-tempered and witty: the English as sportsmanlike, intelligent and conventional: the Italians as artistic, impulsive and passionate: the Jews as shrewd, mercenary and industrious: Americans as industrious, intelligent and materialistic: Chinese as superstitious, sly and conservative: Japanese as intelligent, industrious and progressive: Turks as cruel and very religious.

One illuminating feature in this experiment was the fact that there was less agreement concerning the outstanding characteristics of the Americans who ought to have been known best, as the people who did the experiment were Americans themselves, than upon the characteristics of the Jews, Germans and Italians. It may easily be that the less one's acquaintance with a national group the more is one influenced by a stereotyped attitude in regard to it.

Now stereotypes such as these are undoubtedly ephemeral: they depend on the particular circumstances of the time. One has seen in recent years the rapid change in stereotypes concerning the Russians, the Germans or the French which have followed changes in political alignment. They do, however, serve the purpose of filling a void in our information: but they fill it so loosely that they can be readily exchanged for another stereotype as the occasion requires.

All stereotypes are not quite so superficial and ephemeral. A slightly more stable type is probably to be found in attitudes towards Negroes, though even here very great national differences are observable. In the course of conversation with hundreds of people in France, Lapiere[1] managed to introduce the question, "Would you let a good Negro live at your home?" whenever a suitable occasion presented itself. He found that although there was prejudice against the Negroes among the upper classes (only 8 per cent. of his interviewees answering the question in the affirmative), yet among the middle class 52 per cent., and among the lower classes 86 per cent. were without prejudice. Lapiere then continued his investigation in England, but anticipating a greater prejudice he modified the question to, "Would you let children associate with those of good coloured people?" He found a far higher degree of prejudice, and the class differences tended

[1] Lapiere (10).

to be in the reverse direction from those in France. Of the upper classes 73 per cent. answered the question in the negative, of the middle classes 78 per cent., and of the lower classes 90 per cent. A similar difference was shown between the hotel managers in France and England. In France 24 out of 31 who were asked if they would admit Negroes expressed complete willingness, while in England 16 out of 20 said they would exclude both Negroes and Indians.

Now although it might be possible to change these attitudes by means of propaganda, it would probably be more difficult to do so than, for instance, to change the English attitude towards the Turks from contempt to admiration or *vice versa*. Such stereotypes, whether ephemeral or more stable, arise as a reflection of the cultural pattern both of the people making the generalisations and of those about whom they are made.

Conclusion

At the beginning of this chapter we investigated the concept of the nation and tried to see how it differed from that of the race. We considered various objective criteria, such as language, religion and territory and reached the conclusion that although each was important as an integrating and unifying bond none, not even the possession of territory, was a *sine qua non*. But on the subjective side we were drawn to the perhaps rather vague and unsatisfactory conclusion that the essential element in nationality was the existence of a spirit of national self-consciousness. If that is so it ought to be possible to measure it, but here we found ourselves in a difficulty, for although it might be possible to show cases where the spirit of national self-consciousness did not exist, such information was of a purely negative kind and it was not possible at present accurately to measure the strength of national self-consciousness when it did exist. All that we could say was that in certain specified instances it was certainly strong enough for true nationhood to exist.

Having reached this somewhat unsatisfactory conclusion about the determinants of a nation we next turned to the more important question from the psychological point of view, namely the existence of national character. Here, too, we found that although it has been a common practice to generalise about what are considered to be the typical ways of behaving of people belonging to different nations, yet in order to get any really reliable information a far more detailed analysis had to be made of the conditions of life of those nations than any one has hitherto attempted to make. Thus it was essential to consider the relationship between a particular group of people and the physical features of its geographical environment, the type of country in which it lives and the climate it enjoys in order to see not how these features may have determined the behaviour of the inhabitants of that country but in order to see how the inhabitants have made use of and modified the possibilities with which they have been presented. And in this analysis, too, it was necessary to consider not only the present use of possibilities but also the possibly different uses that might have been made of such possibilities in that nation's history. Secondly it was necessary to consider various economic factors—in particular the density of population, the quality of the population, the relative proportion of the different sexes and the age distribution—for each of these factors must inevitably affect the possibilities which are open to the group and the use which it makes of them. In the third place it was necessary to consider the group's cultural pattern. Which ways of behaving does the group happen to encourage, what are its taboos, its traditions and its customs, and have these been chosen, as Benedict suggests, purely by chance out of all the myriad possible ways of behaving, or, if not, how are they related to the geographical circumstances and the economic conditions in which the group finds itself now and has found itself in the past?

Even if we succeed in obtaining satisfactory answers to these questions, we have still to face four further difficulties. We have to consider the type of political organisation within the group, for on that may well depend which characteristics will be allowed to express themselves and which will be suppressed. We have to consider also the form of class structure in the nation, so that we may obtain some idea of the degree of social homogeneity within it, and so that we may be careful to avoid describing the characteristics of a particular class as if they applied to the nation as a whole in those cases where the social differentiation is marked. Thirdly we need to know the degree of cultural homogeneity as well as social homogeneity within the nation: we need to know in particular its language and religion in order to see how far these tend to unify or to divide the nation. In the fourth place we have to consider the level of maturity which the nation has reached, so as to be on our guard when we are comparing the national characteristics of different groups against regarding the characteristics of two groups as fundamentally different when they may only be a reflection of the stages of maturity which those two groups have attained.

No one has so far made a study of national characteristics on this exhaustive basis. This fact is disappointing. Negative conclusions always are. It need not, however, be disheartening if the difficulties we have mentioned can be used to point in the direction towards which a full and factual analysis of national characteristics will eventually emerge. Ginsberg has already described some of the characteristics of the English and of the Germans in terms of their historical development, and the full field of study is one that would well repay investigation. One might even claim that it is essential for such an investigation to be undertaken for it would be likely to lead to a better understanding of the strengths and weaknesses of ourselves as well as of other nations, of our and their commonest types of stereotyped ideas and behaviour, and of the most fruitful lines to adopt in order to get that real co-operation between nations on which the peace of the world will depend.

REFERENCES

1. Barker, E., *National Character.* London: Methuen, 1927, pp. 288.
2. Benedict, R., *Patterns of Culture.* London: Routledge, 1935, pp. 291.
3. Benedict, R., *Race and Racism.* London: Routledge, 1942, pp. 175.
4. Blackburn, J., *Psychology and the Social Pattern.* International Library of Sociology and Social Reconstruction. London: Kegan Paul, 1945, pp. 157.
5. Forde, D., *Habit, Economy and Society.* London: Methuen, 1934, pp. 500.
6. Ginsberg, M., "National Character and National Sentiment", in *Psychology and Modern Problems* (Ed. J. A. Hadfield). London: Univ. of London Press, 1935, pp. 241.
7. Ginsberg, M., "National Character", *Brit. J. Psychol.*, 1942, *32*, 183-205.
8. Hertz, F., *Nationality in History and Politics.* International Library of Sociology and Social Reconstruction. London: Kegan Paul, 1944, pp. 417.
9. Huntington, E., *Civilisation and Climate.* New Haven, Conn.: Yale Univ. Press, 3rd edn., 1934, pp. 453.
10. Lapiere, R. T., "Race Prejudice: France and England", *Soc. Forces,* 1928, 7, 102-11.
11. Madariaga, S. de, *Englishmen, Frenchmen and Spaniards.* Oxford: University Press, 1937, pp. 256.
12. Mead, M., *The American Character.* London: Pelican Books, 1944, pp. 173.
13. Renan, E., *Qu'est-ce qu'une nation?* Paris: Lévy, 1882, pp. 30.
14. Semple, E., *Influences of Geographic Environment.* London: Constable' 1911, pp. 683.
15. Stout, G. F., *Manual of Psychology.* London: Univ. Tutorial Press, 4th edn. rev., 1929, pp, 680.
16. Blackburn, J., "A Review of Gestalt Psychology", *J. Ment. Sci.*, 1940, *86*, pp. 1-28.
17. Koffka, K., *Principles of Gestalt Psychology.* London: Kegan Paul, 1935, pp. 720.
18. Köhler, W., *Gestalt Psychology.* London: Bell, 1930, pp. 312.

CHAPTER VII

CLASS DIFFERENCES

"My University tailor had a daughter, whose dower he announced as £30,000, and he gave out that none but a gold-tassel should be allowed to cultivate her acquaintance. But the young noblemen never came, and the damsel pined for a couple of years. The father widened the bounds, and gentlemen-commoners were admitted, but still the maiden was unwooed. In another three years the suffrage was extended to all members of Christ Church. There may have been wooers now, but no winners. Five years more and the maiden still sat at her window unclaimed. For another five years the ninth part of a man held out resolutely, but by that time youth was gone, and the daughter so long a prisoner was glad to accept the hand of an aspiring cheesemonger."—*The Habits of Good Society: A Handbook of Etiquette for Ladies and Gentlemen.* London: Virtue and Co., 1874.

Introduction

MacIver[1] defines a social class as, "any portion of a community which is marked off from the rest, not by limitations arising out of language, locality, function, or specialisation, but primarily by social status". This definition at once raises the question what is social status and how is it determined? Is social status determined by sex, age, relationship, birth, occupation, education or by the possession of more or less wealth or intelligence? Or does it depend on a way of life and outlook, a feeling of belongingness to one group rather than to another? Or does it consist of a complicated mixture of these objective and subjective criteria?

Most writers on social class have pointed out the distinction between horizontal and vertical groups. In both cases the different groups may be superficially distinguished from one another on the grounds that they are formed by people having similar interests, outlooks and ways of behaving. In both cases there may sometimes be practically no mobility between the different groups, and at other times a considerable amount. Examples of horizontal groups are literary or musical associations, occupational groups and so on. Between literary and musical associations there may often be a considerable degree of mobility, many members of the one also being members of the other, but between

[1] MacIver (13) pp. 166–7.

occupational groups, and other groups of a similar kind, the mobility may be negligible or non-existent. Groups of Scotsmen and of Welshmen living in London, for instance, often form highly integrated groups, yet there is no mobility between the one group and the other. All these are horizontal groups.

In vertical groups there is a social recognition of the fact that some groups are superior to others. The fact of social recognition is important, for, as Marshall[1] points out, the relationship of superiority and inferiority is not based merely on natural differences. The natural superiority of the leader over the follower is not of the nature of social class: it requires social recognition to make it so.

Just as the ease of movement between different horizontal groups varies according to the horizontal distance (that is to say the similarity or difference in outlook, interest and behaviour) between the groups, and according to the structure of the society concerned, so also the mobility between different vertical groups varies according to the vertical distance between them, and according to the cultural pattern. Most societies appear to have possessed some form of class system, though the degree of mobility between the different classes has varied enormously from the rigidity of a caste system (the limiting form of class) to the ease of movement to be seen, for example, in modern Czecho-Slovakia.

In democratic countries at the present day legal barriers to movement from one class to another (both within a single generation and between successive generations) have been largely or entirely removed. Yet it remains a fact that social barriers have not all disappeared. Legal equality, as Aron[2] points out, makes movement easier, but it does not by itself increase or hasten it. For social mobility is largely independent of laws: it rests on economic conditions, on social prejudices, on the prestige which is attributed to different occupations, and on the values which are attached by society to different elements within its own pattern of culture.

Objective Determinants of Class Differences

MacIver's definition of a social class eliminates those portions, of a community which are marked off from the rest by limitations arising out of language, locality, function or specialisation. It is true that these and other criteria of the same kind, e.g. sex, age,

[1] Marshall (14) p. 58. [2] Aron (1) p. 13.

birth, race, wealth, education and so on have often been used by societies to mark off different levels of superiority or inferiority. Yet MacIver holds that they are in themselves insufficient to establish cohesive groups.

(a) *Sex*

This argument is more readily acceptable in the case of some of the characteristics mentioned than it is in the case of others. Many societies attribute a position of greater prestige to men than they do to women. Women often form a group intermediate in prestige between men and children. In such societies if we consider merely the three groups, men, women and children, and their reciprocal behaviour towards one another, we can pick out the same kind of deferential or condescending behaviour as we can see in the behaviour of one social class towards another. Yet to group together all women as a social class distinct from all men or all children is immediately to introduce a false sociological note. Social class cuts across sexual divisions and it cuts across age divisions. As Marshall[1] points out, "A group of men endowed with rank or titles for which only men are eligible may play an important part in the formation of Classes, but it is not a Social Class. For our purposes the significant group created by the English Peerage is not composed of the men who sit in the House of Lords, but includes their families and relatives of both sexes through several degrees of affinity. The 'professional Class' is not co-terminous with the members of the professions. Wherever women take their social position from their fathers and husbands, Social Class is effectively determined by the status of men, but it is not composed of men. We cannot accept the familiar generalisation that Class tends to endogamy and also speak, in a literal sense, of a military Class, since there are obvious difficulties in the way of making an army endogamous."

(b) *Age*

Sex, therefore, does not determine social class. Nor does age. Many societies base a hierarchical system on age, endowing the older people with more prestige. The differentiation of children from adults has already been mentioned in the last paragraph. But even in our own society there is a further differentiation among adults themselves. At any rate up to the age of 65, and sometimes far beyond that age, an adult's prestige, social pre-

[1] Marshall (14) p. 63.

cedence and so on is largely determined by the number of years he happens to have lived. And in other societies than our own the status accorded to a person, the social rights he enjoys, the power with which he is endowed, depend more on the age group to which he belongs than on any other single factor.

Yet although age may sometimes be employed as the principal method of determining the social hierarchy, it is as false to regard it as the principal or sole determiner of social class in our own society as it is to regard sexual differentiation in the same way. Although one may find certain things in common with people who are of the same age group as oneself, yet social class cuts across this division. Furthermore, as Marshall[1] points out, in our own society social class is thought of as having a certain amount of stability and permanence. And it is this essential feature which distinguishes it from a system of seniority. A person will remain in the social class into which he is born unless he himself, or someone on his behalf, takes steps to push him up or down. But one cannot help growing older, and the mobility from the lower age groups to the higher, with its concomitant advantages in prestige, is therefore automatic. Furthermore, the mobility between age groups works in only one direction: one gets older, one does not (unfortunately) get younger. In social class, however, there is the possibility of movement downwards as well as upwards, even though in our own society at the present day the stronger pressure is upwards, and more trouble is taken both by individuals and by social groups to prevent themselves or their members from falling into a group of lower social status than to prevent them from rising.

(c) *Conquest*

Before turning to birth and relationship as objective criteria determining class, one word needs to be said about conquest. Obviously where one group is conquered by another in warfare, the conquering group will gain prestige by their conquest, and this prestige may be extended from their military prowess to cover other aspects of their behaviour as well. The conquering group will be more likely to form a caste than a class, at first at any rate, though later, when there has been a certain softening in the rules of non-fraternisation, and consequently a certain amount of mixture between the two groups, the descendants of the conquerors may form a superior class to the descendants of

[1] Marshall (14) p. 59.

the conquered. Although conquest has been the cause of the development of classes in some communities, it is doubtful whether even the Norman Conquest determined the development of classes in this country.

(d) *Birth and Relationship*

In birth and relationship we reach much more substantial criteria. Hereditary factors are of especial importance in primitive communities, and they remain so in western civilisation. Castes of kings, chiefs, nobles and priests are largely determined by birth, and when once a change appears from the rigidity and immobility of a caste system to the greater mobility of a class system, birth tends to remain as an important factor in determining the class to which a person belongs. It remains so to this day in this country. A child born of aristocratic parents, or born into one of the families which have been associated with ruling this country for generations, is at first, at any rate, accepted as belonging to a superior social class. We have already mentioned in the quotation from Marshall that even where men only are eligible for a particular rank or title their social status is extended to include their families and relatives of both sexes through several degrees of affinity. This leads Marshall to follow Schumpeter[1] in saying that the true unit of social class is the family. Even the new-born infant belongs to the social class of his parents.

Yet however important birth and relationship may be as factors in the determination of a person's social class, they are not an essential nor the sole determiners of it. Whether a person continues to be accepted as belonging to a particular social class will depend on his way of behaving later on, what kind of education he is given, what occupation he takes up and what skill he employs in looking after his money. Even, however, if he loses his money and takes a job not often adopted by members of his class he may still retain the prestige of a higher social class himself, though it is less likely that his children will. But birth by itself is not, we repeat, a sufficient determiner: there are many accepted members of the higher social classes who come from neither aristocratic nor from "good" families.

(e) *Occupation*

Is type of occupation the sole determiner of social class? There is undoubtedly an association between social class and type

[1] Schumpeter (16) p. 12.

of occupation. It is known,[1] for example, that most children enter the same occupation as their fathers—partly no doubt because of the relative immobility of labour, and partly because fathers know of openings in their own lines of work, and know more about what the occupation entails, than they do about other types of work. Thus it has been shown that nearly 50 per cent. of the sons of clerical workers enter clerical occupations, and over 60 per cent. of the sons of textile workers enter the textile industry. One estimate is that the pull of the father's occupation compared with the pull of any other occupation is as 3 to 1. Ginsberg's[2] analysis of 1,268 British subjects admitted to Lincoln's Inn between 1886 and 1927 showed that 75 per cent. of them belonged to the upper and middle class of the Registrar General. Only during the last four years of the period investigated did the admissions of sons of wage earners amount to more than 1 per cent. of the total, and even then they were less than 2 per cent.

In relation to occupation it has been argued[3] that a determinant of class is to be found in (a) the differentiation into those who give orders and those who receive them, and (b) the inferiority of the seller in the seller-buyer relationship. Both of these depend on occupation but neither is, of course, an explanation of class difference. A policeman or a railway guard often gives orders which are obeyed by people who would consider themselves, and who would be considered by him, to be of a superior social standing. Again, in conditions of scarcity, or in certain types of organisation of production or distribution, it is the seller who is in the most advantageous position and tends to gain prestige thereby. But even in other, more usual, economic conditions the seller is not always in an inferior social position, for instance if a penurious aristocrat peddles silk stockings or other wares in a middle middle-class district his "cultured" voice gains for him immediate prestige in the ears of the householder.[4] Consequently we see that differences in occupation are not a complete explanation of the fact of class differences. As Tawney[5] has pointed out, stockbrokers, barristers, doctors, miners, railway men and cotton spinners form half a dozen occupations, but they do not form half a dozen different classes. Similarly, postmen, bricklayers and engineers pursue sharply contrasted occupations and often

[1] Carr Saunders and Caradog Jones (2) pp. 132–3.
[2] Ginsberg (8) pp. 170–71.
[3] See Thouless (18) pp. 336–41 or Pear (15).
[4] See Wootton (22).
[5] Tawney (17) pp. 68–9.

have divergent economic interests, but they are not distinguished from one another by the kind of differences associated in common opinion with differences between classes. It would be truer to say, therefore, that class tends to determine occupation rather than that occupation tends to determine class.

(f) *Education*

Next let us consider the relationship between education and class. The State system of education is avoided in this country by a minority whose parents possess wealth, birth and/or the right occupation. By doing so a considerable advantage is obtained, if one wants to obtain a well-paid and influential occupation. Public schools give special advantage, but most of the better paid occupations are not open to people who have not had some form of secondary education. In 1935[1] less than 12 per cent. of the children leaving elementary schools proceeded to a secondary school, and of these 90 per cent. gained free or partly free places as the result of an examination which was taken at the age of 11+. If the poor child is to get any advanced education, therefore, he will suffer from the pressure of the examination system from a very early age, and even if he is successful in obtaining a scholarship which gives him a free place in a secondary school his parents will nevertheless have to make some financial sacrifice, for he will be postponing the time when he enters gainful employment. Unless he is very fortunate and/or clever, therefore, the child who goes to an elementary school will be in a position of considerable disadvantage compared with the child who takes the first stages of his education outside the State system.[2]

The chance of an elementary school child reaching a university is still more remote. The odds lengthen from the 17 to 2 against his proceeding to a secondary school to about 250 to 1 against his going to a university. *Prima facie* it would appear to be unlikely that *all* the 250 unsuccessful ex-elementary school children would fail to profit by a university education, and attendance at a university is considerably more of a guarantee of a privileged job than attendance at a secondary school, or even at a Public school. Type of education, therefore, is more than occupation a determinant of social class, but again it is not the sole explanation. Class also determines education.

[1] Carr Saunders and Caradog Jones (2) pp. 119–20.

[2] The position will be greatly improved when the provisions of the Education Act of 1944 are in full working order.

(g) *Wealth*

Can wealth be the explanation of social class differences? On wealth will largely depend the type of education a person gets, and the kind of occupation which he subsequently enters. Many people have held that wealth is the sole determinant of social class. I have myself no doubt that it is a far more important single determiner than any of the other factors already considered. I do not, however, believe that it is sufficient as a sole determiner. The fact of social class is more complex than that.

Carr-Saunders and Caradog Jones[1] argue that it cannot be used as a determinant because it is continuously graded. Is it not completely arbitrary and psychologically untrue to say that a person earning £249 a year belongs to one social class while a person earning £251 a year belongs to a higher one?[2] Besides this argument there is the fact that people of the recognised upper social classes in straightened circumstances still retain much of the prestige of their class.

(h) *Intelligence*

Finally, as the last of our objective criteria, let us consider intelligence as a determinant of class. Investigations in both England and America have revealed that there is a positive association between intelligence test scores and occupational class. This has been found both when the tests have been applied to the children of parents who come from different occupational classes, and when they have been applied to the parents themselves. An example of the latter type of investigation is the result of the tests which were given to American recruits in the war of 1914–18. It was found[3] that the average score in different occupations could be differentiated, and that it ranged from that of labourers and factory hands at one end of the scale to that of school teachers and engineer officers at the other.

In Cattell's investigation[4] in this country tests were applied to a large number of people in different occupations. Cattell found the average levels in different occupations were as shown in Table 13.

Nevertheless, although the average I.Q. in different occupa-

[1] Carr Saunders and Caradog Jones (2) p. 66.

[2] Class divisions of this kind are made in market research investigations. There is some justification for it here, for the investigators are primarily interested in questions of spending power.

[3] Yoakum and Yerkes (21) p. 198. [4] Cattell (3) p. 21.

		I.Q.
30	Secondary school teachers	151
25	Physicians and surgeons	146.5
50	Central school teachers	145
20	Civil engineers	142
18	Mechanical engineers	140
90	Elementary school teachers	137
15	General managers in business	137
57	Shorthand typists	129
54	Commercial clerks	127
14	Typists	126
24	Commercial travellers	123
250	Nurses (probationer)	122
19	Telephone operators	120
52	Precision fitters	114
14	Coach body builders	106
20	Sheet metal workers	102
12	Shop assistants	99
33	Carpenters	98
31	Cabinet makers	97
33	Machine operators	96
24	Coach trimmers	96
12	Hairdressers	89
16	Upholsterers	87
12	Welders	87
19	Factory packers and sorters	78

Table 13. Intelligence Quotient and Occupation. (From Cattell (3), p. 21.)

tions showed a continuous decline, there was an enormous overlap between the different groups. For instance, the highest scores of the factory packers and sorters whose average I.Q. was 78 were greater than the lowest scores of the sheet metal workers, whose average was 102: the highest scores of the upholsterers whose average was 87 was greater than the lowest scores of the precision fitters, whose average was 114: the highest scores of the machine operators whose average was 96 was greater than the lowest scores of the typists, whose average was 126: and the highest scores among the shop assistants whose average was 99 was greater than that of the lowest general managers of businesses, whose average was 137.

Recently Himmelweit and Whitfield published an analysis of over 5000 intelligence test scores obtained from Army recruits at training centres during the early part of the war. The mean value and standard deviation of the test scores were calculated for each of 39 main occupations (that is to say those occupations which were represented by not less than 50 individuals). Each occupation's mean score was compared with every other occupa-

Occupation	*No. tested*	*Mean test score*	*Standard deviation*	*No. of occupations with higher mean not significantly different*	*No. of occupations with lower mean not significantly different*
Schoolmaster }[1]	92 }	40.65 }	7.46 }	0	1
Teacher	109	39.68	7.22	1	0
Student	237	37.66	7.72	0	0
Bank Clerk	104	34.97	7.83	0	0
Railway Clerk	180	31.79	7.72	0	0
Clerk	197	28.93	9.77	0	1
Messenger	54	25.80	7.75	0	2
Electrician	50	25.78	10.13	2	3
Baker	63	23.30	7.22	2	3
Shop Assistant	118	22.76	8.89	2	4
Shop Manager	136	22.76	7.65	3	3
Grocer	57	21.51	8.15	3	16
Fitter	70	20.48	8.24	3	14
Wood Machinist	110	20.33	7.80	2	10
Shoe Operative	52	20.33	7.75	5	13
Bus Conductor	99	20.15	7.75	4	11
Postman	61	19.21	9.62	5	16
Mechanic	56	19.16	7.84	6	13
Fitter's Mate	50	19.05	8.94	7	16
Van Driver	112	18.88	7.81	8	10
Carpenter	60	18.88	8.94	9	14
Salesman	63	18.78	8.62	10	13
Storekeeper	70	18.45	8.25	11	12
Motor Driver	229	18.31	7.88	11	8
Roundsman	119	18.18	7.59	12	10
Warehouseman	219	18.16	8.64	11	8
Builder's Labourer	58	17.98	7.64	15	8
Lorry Driver	416	17.94	7.56	11	6
Butcher	64	17.73	8.02	16	8
Driver's Mate	79	16.73	8.63	13	7
Painter	98	16.38	8.46	12	6
Driver	133	16.32	8.49	12	5
Porter	68	16.31	8.64	15	4
Gardener	139	16.25	8.19	11	3
Railway Porter	59	15.76	8.31	14	2
Miner	60	14.90	8.31	7	3
Cleaner	50	14.68	8.67	8	2
Labourer	1113	13.40	7.80	2	1
Carter	50	12.04	8.56	3	0

TABLE 14. Test Scores of Thirty-Nine Occupations. (From Himmelweit and Whitfield (12) p. 226.)

[1] Himmelweit and Whitfield give these two occupations separately following on their decision to concern themselves merely with the description of his occupation given by each person. They believe, however, that it would probably be better to regard these two as a single group.

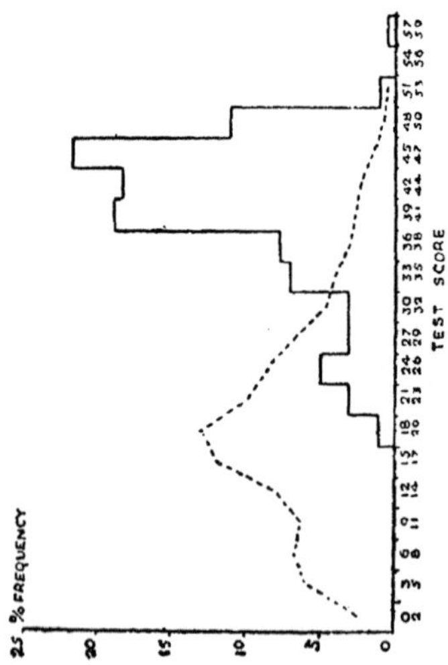

Schoolmasters and Teachers.

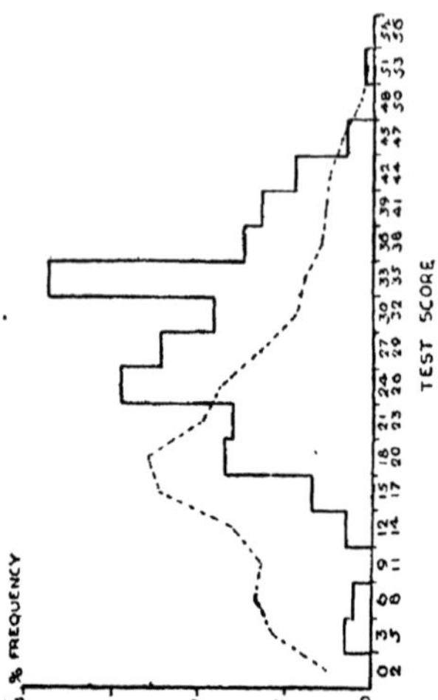

Clerks.

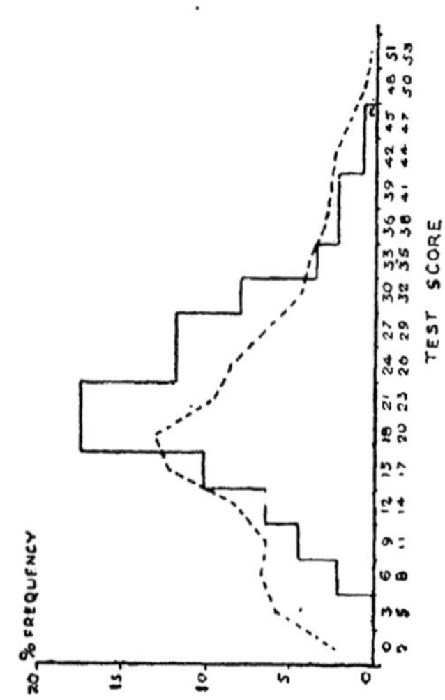

Shop Managers.

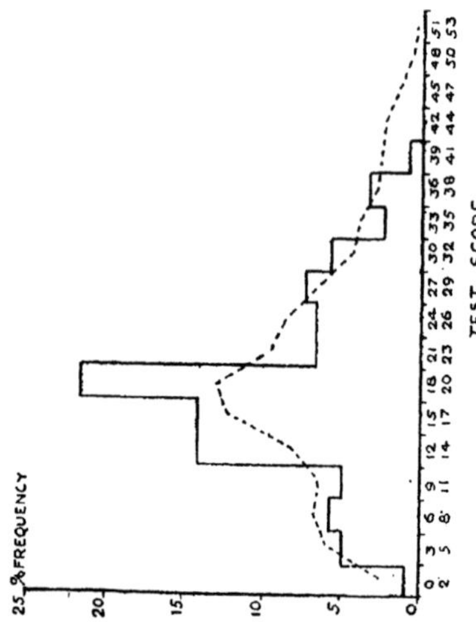

Fitters and Fitters' Mates.

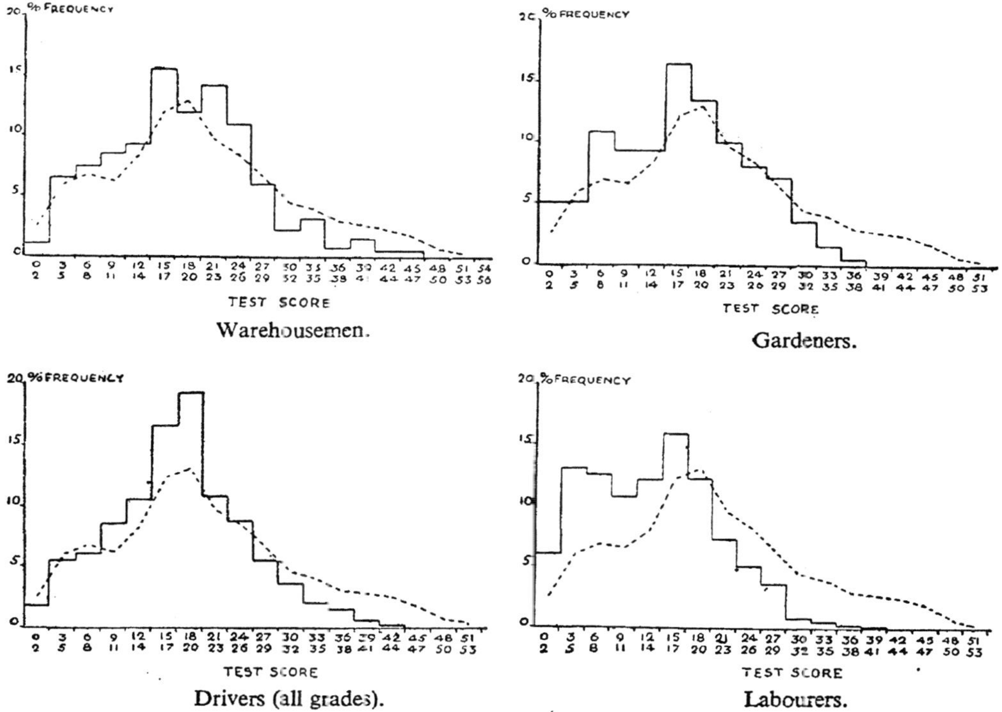

FIG. 6. Test score distributions for certain occupations. (From Himmelweit and Whitfield (12) p. 225.)

tion's mean score to see whether the difference between them represented a real difference in intelligence test score, or whether the difference was merely such as might be expected by chance. The results are shown in Table 14.

Himmelweit and Whitfield also published the test score distributions for eight of the occupations. These may be seen in Fig. 6.

In each figure the histogram shows the occupational distribution and the dotted line the distribution in all the 39 occupations together. For comparative purposes both the histogram and the dotted line are based on the percentage of individuals who obtained each score.

In this investigation, as in the other two we have mentioned, we see again that the average score in different occupations diminishes as we pass from the higher to the lower occupational groups. There are, however, certain points that need to be remembered about these investigations. So far as the recruits are concerned (a) the analysis of occupations was based on the information which the recruits themselves provided: the accuracy of their statements was not checked, and (b) the medically unfit, volunteers for other services, and those (in certain occupations) reserved as key men were not included. In addition to these points there is in none of the investigations we have mentioned any information available about the length of time a person had been employed in a given occupation, nor about his proficiency in it.

All these investigations, therefore, show an association between the level of intelligence test score and type of occupation, but they also show that the range of intelligence within any occupation is large. Himmelweit and Whitfield's investigation indicated that the middle range of occupations showed much overlapping in intelligence levels, and that they included representatives of each intelligence level in approximately the same proportions as the general population. It seems, therefore, that a wide range of intelligence is tolerated in most occupations.

A somewhat similar picture emerges when we consider the results of tests applied not to adults but to the children of parents in different occupations. An example of this is to be found in the work of Duff and Thomson[1] who gave intelligence tests to over 13,000 children in elementary and secondary schools in Northumberland, and grouped the fathers' occupations into one

[1] Duff and Thomson (6).

of four classes—brainwork, skilled handwork, semi-skilled handwork, and unskilled. They found that the average score increased in the higher classes, as can be seen in Fig. 7.

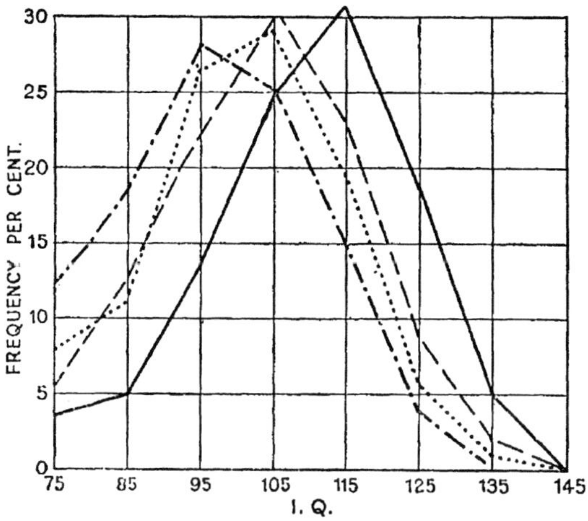

——— = Brainwork. — — — — = Skilled Handwork.
. . . . . = Semi-skilled Handwork. —·—·—· = Unskilled.

FIG. 7. Intelligence Quotient and Social Class. (From Waddington (19) p.349.) (Data from Duff and Thomson (6).)

Here again there is an enormous overlap between the different classes, and since there are a greater number of people in the lower than in the higher occupational classes the actual number of people above the average may be higher in the lower classes than it is in the upper.[1]

In addition to this we must remember that the occupational classification of fathers overlooks the fact that children inherit half their characteristics from their mothers. Unless therefore occupational classes are completely endogamous (which is by no means the case—particularly in so far as neighbouring classes are concerned), or unless we have some accurate measurement of the extent of the endogamy and exogamy (which we do not at present possess), we cannot expect to make any reliable deductions about the extent of the genetic differences in intelligence between different occupational classes from investigations such as this. This is particularly important to remember when the differences are as slight as they are between skilled and semi-skilled workers.

To put the point in a different way, we could say that the

[1] See Waddington (19) p. 347.

differences in intelligence between different occupational classes found in this investigation *must be due to environmental causes* if it were the fact that there was unrestricted exogamy between the different occupational classes—if a brainworker were as likely to choose a mate from the unskilled group as from any of the other three groups, etc. In fact this is not so, but the greater the lapses from complete endogamy within the different occupational classes the more have we to attribute any differences in the intelligence levels to environmental influences, and as we do not know accurately to what extent endogamy and exogamy occur it is impossible to know how to interpret such differences as are found.

Yet, as we saw in Chapter III, some psychologists have held that there is a difference in innate intelligence between the classes, and when this is combined with the fact that the net reproduction rate is at present higher in the lower classes than it is in the higher, it gives rise to the view—put forward by Cattell[1] and others—that we are racing down a slippery slope towards feeble-mindedness.

Fisher[2] holds that this must of necessity be the case in any society which uses the acquisition of wealth as one of the main steps by which a person can rise to power or eminence. He argues that in such communities the people at the top will be composed of two classes—the able and the infertile. The infertile will rise because a larger proportion of the family's resources will be available for spending on their education than will be the case if they have many brothers and sisters. Thus these two classes at the top will tend to inter-marry. The intelligent will marry the infertile and will not maintain themselves in consequence—for Fisher believes (and produces evidence in support of his hypothesis) that there is a considerable hereditary factor in fertility. Again, if an intelligent and/or infertile man marries an heiress there is a high probability that his family will become extinct, for heiresses have a higher degree of infertility than the average, since they come from families in which there are no sons. Thus Fisher holds that these facts are likely to bring about the decay of our civilisation, just as—he thinks—they have brought about the decay of previous civilisations.

Fisher's argument has, however, been criticised, particularly by Charles[3] and by Haldane.[4] In the first place it is argued that

[1] Cattell (4).
[2] Fisher (7).
[3] Charles (5) esp. pp. 126–43.
[4] Haldane (10) pp. 120–27.

infertility is a far more important determinant of the rise to the upper classes than is intelligence. We have already seen that there are grounds for regarding with suspicion the different intelligence test scores of different classes, and if this is so then any differential fertility that may be found will not necessarily eliminate the intelligent. Secondly, such differences in intelligence as are found are based on the results of intelligence tests, and we have already put forward in Chapters III and V a number of reasons why it may be that intelligence tests do not accurately measure the amount of a person's innate intelligence. Thirdly, recent investigations have shown that the differential fertility between different classes is becoming much smaller as knowledge about birth control methods has spread to the lower classes. And finally other factors than intelligence, some desirable and others undesirable, may be of as great or greater importance in our society in determining the struggle for success—factors like energy, persistence, stability, selfishness and dishonesty which are not measured by intelligence tests. As Himmelweit and Whitfield [1] put it, "A person of average intelligence has a wide choice of occupations open to him as far as intelligence requirements are concerned. For vocational guidance or selection this means that, except at the extremes of the intelligence scale, factors other than intelligence must be considered in fitting individuals to their most suitable occupations."

Subjective Determinants of Class Differences

If none of the objective criteria we have considered is sufficient to explain the phenomenon of social class, can we obtain any more light on the subject by turning to subjective criteria? Is a person's social class determined by his feeling of belongingness to a particular class, by sharing in common with other people a particular kind and degree of class consciousness?

At the beginning of this chapter we mentioned the distinction between vertical and horizontal groups, and pointed out that social classes were one form of vertical grouping. Before this point is elaborated any further it will be as well to discuss the relationship between group consciousness and class consciousness. Groups of Scotsmen and Welshmen living in London were taken as an example of horizontal grouping: for although the Scotsmen may regard themselves as superior to the Welshmen, the Welshmen are not prepared docilely to accept a position of inferiority,

[1] Himmelweit and Whitfield (12) p. 224.

and instead of regarding themselves as inferior to the Scots tend to regard themselves as superior to them. Both groups may possess, therefore, a high degree of group consciousness. But this is not the same thing as class consciousness. Class consciousness involves a complementary attitude between the groups: one group consciously regards itself as superior to a second, and the second acknowledges the superiority (in fact, even if not in theory) of the first.[1]

Group consciousness is often intensified or even artificially created by groups adopting some definite name, a formal organisation, a chairman, a secretary, a constitution, an esoteric terminology, elaborate initiation rites, and special ways of behaving, eating, speaking and so on.[2] These are seen in their most extreme form (with the added entrancing touch of mystery about them) in the Ku Klux Klan, the Royal and Antediluvian Order of Buffaloes, and Freemasonry, but they occur in a much less extreme form in the ordinary organisation and rules of procedure of many associations, clubs and societies.

But group consciousness is not the same thing as class consciousness. There is a complementary attitude of superiority by one group towards a second and of inferiority by the second towards the first in class consciousness. It is debatable whether, from the psychological point of view, all cases in which these complementary attitudes exist may be regarded as cases of class consciousness or not. An attitude of superiority may be adopted by the old towards the younger, or by men towards women, and this attitude may find its counterpart in an attitude of inferiority adopted by the younger or the women towards the older or the men. Yet we have already seen that age and sex do not form social classes. Is it fair, then, to regard the attitudes adopted by the members of the different groups towards one another as the same thing as that which exists between the members of different social classes? It is possible that the attitudes *are* the same, though the values towards which the attitudes are oriented varies in different cases.

If this is so, then we can say that class consciousness will tend to start in relation to the possession or non-possession of any things on which the social group places value. It will tend to

[1] See Ginsberg (9) pp. 160–61.

[2] Codes of manners, ways of behaving, eating and speaking and so on may become the most important immediate (though superficial) means of identifying a person's class once the idea of class differences and the fact of class consciousness have made their appearance.

start in relation to the possession of wealth in communities in which the acquisition of wealth is a major goal, in relation to occupation where some occupations are given greater social esteem than others, to education where some kinds of education are in a privileged position, and so on. From these factual conditions of superiority and inferiority the attitude will become generalised to cover other attributes, so that one group is considered to be superior to a second not only in the fact, let us say, that it possesses more wealth, but also, by generalisation from the fact of wealth, because it is held to possess a host of other characteristics which may be entirely unconnected with the possession of wealth, such as virtue or ability or dependability and so on.

Furthermore we can say that psychologically the same phenomenon may occur within a group *of the same social class.*[1] Thus, for instance, at some schools which place a high value on athletics those who shine at athletics will tend to form a class with an attitude of superiority over those who are not so able at sports as themselves. And the poorer games players will reciprocate with an attitude of inferiority towards the athletes. Both groups, too, will extend the acknowledged superiority of the athletes in sports to cover other aspects of their behaviour as well. In a similar way class consciousness may arise in a school which places its main value on intellectual prowess.

Class consciousness arises, therefore, as a result of the recognition by two groups that one is superior to the other in some respect having social value, and this acknowledged and factual superiority tends then to become extended to cover other aspects of behaviour as well. Social *snobbery* is an expression of the attempt by those in the acknowledged inferior class to raise themselves into a superior one and to prevent themselves from falling into a lower class. Although both snobbery and class consciousness are found between classes which are widely separated in distance, they are most frequent and most influential when the classes are close to one another, especially when there is in those circumstances a very real possibility of movement from the lower to the upper class or of falling from the upper to the lower. Thus although advertisements showing members of the aristocracy using a particular form of face cream or a particular brand of whisky appear to have a wide appeal, yet the circulation of those papers which devote a large proportion of their contents to

[1] Psychologically it is the same although sociologically it is very different.

photographs of the ladies and gentlemen of society is said to be largest among those who are not quite high enough themselves to be the objects of the photographs, but who with a little luck might be, and who would certainly like to be. Athletic snobbery and intellectual snobbery have much in common with social snobbery: they differ chiefly in the values on which they are based.

Class consciousness, therefore, not only affects one's attitude towards other classes and their attitude towards oneself, it also, when it becomes snobbery, affects one's behaviour. This competitive emulation is also to be found very strongly in what has been called the psychology of suburbia. An important goal here is to be at least as good as one's neighbours. It is the neighbours (or the standards of the locality) that set the goals. It is important, not to buy a car that is as good as that of the millionaire in Park Lane, but to get one which is as good as or, if possible, a little better than one's next door neighbour's. The important thing is to behave and to speak "correctly"—according to the standards of the neighbourhood. It is perhaps ironical that some of these carefully "correct" ways of speaking, holding the tea cup and so on are frequently used by the class a little higher in the social scale as an immediate indication of the fact that the person is not "from the top drawer".

The attempt to raise oneself in social class, or at any rate to prevent oneself from falling, which are so characteristic of Western civilisation, results in a constant upward pressure. There is a "positive valence" towards the social classes above oneself and a "negative valence" away from those beneath. Illustrations of the way in which these tendencies manifest themselves are given by Harrisson.[1] "All the time in our work," he writes, "we note people tending to talk about themselves as if they are 'better-off' than they really are. For instance, more people will say they read *The Times* than could be true, and nearly twice as many people will say they go to church regularly than in fact can do so. Many difficulties in opinion sampling come from the tendency of people to give what they consider 'respectable' answers, and the consequent difficulty of getting people to give accurate answers, even about their own incomes or how often they have a bath."

"The idea of looking upwards and working upwards in a clear-cut form," Harrisson continues,[2] "first impressed itself on

[1] Harrisson (11) p. 152. [2] Harrisson (11) p. 152.

me at my public school, where there was an elaborate system of privilege and caste, mainly based on the length of time you had been there, but also on how good you were at games. The speed at which you could move downstairs, which waistcoat button could be undone, which hand you could put in your pocket, what cereals could be eaten at breakfast, where you could walk, a hundred habits, were determined entirely in this way.

"After that, I had occasion to spend two years among Melanesians in the New Hebrides and I found very much the same state of affairs. There, in Malekula and Santo tribes, the social pyramid is distinct, based on the number of pigs you can collect and kill. Broadly, each time you kill one hundred pigs, you go up one rung on the social ladder. You gain a few insignificant privileges, much prestige. In the end you reach so high that that you are symbolically off the top of the ladder, earning the title of Hawk and (if you wish) walking about flapping your hands up and down as if in flight."

Bringing the discussion back to England Harrisson continues by describing the graded rites known as "raising" in the Royal and Antediluvian Order of Buffaloes. "The main ritual of the Buffs is around this Raising. One of the dominant motives running through the Order is that of advancement to a higher degree. There are four degrees, or grades, each with its own special medals, sashes, and regalia. Advancement to a higher grade is dependent upon length of membership, attendance at Lodge (branch) nights, number and variety of offices held, and the passing of an examination in the ritual of the degree to which the candidate is seeking admittance; the candidate's Lodge must vote on his fitness for the advancement."[1] After describing the rites Harrisson continues,[2] "Such rites give the participants privileges and badges for ritual occasions, only shown in public at funeral ceremonies or important civic processions. This process of taking small steps upwards in life is probably more important than has generally been appreciated. The steps can be satisfactory even if symbolic, or even if secret."

Stage to stage is the British climb. "At every stage away from the ordinary, first as a shop steward, then a minor Trades Union official, then Ward Councillor, Alderman, Mayor, Union Secretary, M.P., Privy Councillor, Cabinet Minister, the leader changes his accent and appearance almost imperceptibly, gets

[1] Harrisson (11) p. 153. [2] Harrisson (11) p. 155.

farther away from the place where he began—has less and less contact with the masses, has a better house."[1], [2]

THE PATTERN APPROACH

But however characteristic class consciousness may be of the attitude of the members of different classes towards one another, class consciousness by itself cannot be regarded as the sole determinant of class. Class can exist without class consciousness. As Marshall[3] points out, "The new-born infant is not class-conscious and is, socially speaking, intrinsically nothing, and yet there is no doubt that it belongs to a Class and that this fact determines its 'Lebenschance'." Marshall elaborates this view. He regards a social class as a group of people with similar social chances rather than as a group with similar external or internal attributes. "The essence of Social Class is the way a man is treated by his fellows (and, reciprocally, the way he treats them), not the qualities or the possessions which cause that treatment. It would be possible, and perhaps useful, to group people simply in terms of their attributes, without asking how those attributes affected their social relations, but the result would be a study of social types, not of Social Classes.

"Social recognition is, therefore, an important factor even though, in the great majority of cases, a man's Social Class is indubitably determined by the circumstances of his life."[4]

Yet on what does this social recognition rest fundamentally What is the essential determinant of the way a man is treated by his fellows and of the way he treats them? Since none of the objective nor subjective qualities we have considered may be regarded as a satisfactory and sufficient criterion, perhaps the explanation is to be found in a combination of them all, a pattern in which for different classes the different elements play a rôle of varying importance. Perhaps social class differences consist of constellations of factors, constellations which differ for the different social classes. I believe myself that this is the most fruitful line of investigation, though up to the present time it is a line which has hardly been developed at all. A first attempt along it has been made by a group of students directed by Wootton.[5] They define social classes as, "groups of people who

[1] Harrisson (11) p. 159.
[2] For a further discussion of subjective factors the reader should refer to Pear (23).
[3] Marshall (14) p. 64.
[4] Marshall (14) p. 60.
[5] Wootton (20).

mingle with each other in normal social intercourse (e.g. entertain each other in their homes without restraint) but who are conventionally debarred from so mingling with other groups who are socially inferior or superior according to a generally recognised hierarchy."[1] This definition is a behavioural one: it rests on the way in which a man is treated by his fellows, and the way in which his fellows expect him to treat them. There is no reason why it could not be extended to include subjective criteria based on the way in which he feels about them. The investigators then proceeded to examine the characteristics which (in part at any rate) determine this behaviour among the ruling classes in this country.

Taking members of the Government, M.P.'s, Judges and Magistrates, Entrants to the Civil Service Administrative Grade, and the top two or three officers of each Civil Service Department, as forming part of the ruling class in this country the investigators studied their background patterns and found that these fell into the three well-marked types shown in Table 15.

	Type A	*Type B*	*Type C*
Family	Titled	Professional	Not stated
Education	Public School and Oxford	Grammar School and University	Elementary
Occupation	None	Professional	Manual: Trade Union official
Clubs	Carlton and Brooks	Athenæum	Not stated
Recreations	Shooting	Golf	Walking

TABLE 15. (From Wootton (20).)

Among Cabinet Ministers there was a high proportion of Type A: among leading Civil Servants there were variations between Types A and B, with a high proportion of Type B. Judges showed Types A and B. Type C was peculiar to Labour M.P.'s, one-third of them, at the time of the investigation, being of this type. Twenty-one per cent. of them also possessed the common characteristic of being Trade Union leaders. It will therefore be seen that Type C is very unlikely to be found among the ruling classes, except in positions of power to which an individual has been *elected*. It therefore follows that people possessing a background pattern of Type A or Type B are in a privileged position: they are most often found in positions of power, and yet

[1] Wootton (20).

they are by no means typical of an average group of the population. It is true that not all of those who possess a Type A or Type B background are in fact among our rulers, but they have a much better chance of becoming so than do those who do not possess this background. A person possessing a background pattern of poverty, lowly birth, humble occupation and only rudimentary education belongs to a class which is excluded. The most important single determinant in these patterns is the possession of, or lack of, wealth: yet this is not the whole explanation.

A more detailed and extended study of pattern backgrounds would be extremely valuable. Through such a study one might obtain an adequate picture not only of the features which mainly influence entrance into the ruling class, but also of the factors on which differentiation between different social classes in different parts of the country depend. A subsequent extension of such work to other countries, so that the class differentiation in this country could be compared with that in other countries, in particular the United States, would be most illuminating.

REFERENCES

1. Aron, R., *Classes Moyennes.* Paris: Alcan, 1939, pp. 354.
2. Carr Saunders, A. M., and Jones, D. C., *Social Structure of England and Wales.* Oxford: University Press, 2nd edn., 1937, pp. 235.
3. Cattell, R. B., "Occupational norms of Intelligence", *Brit. J. Psychol.*, 1934, *25*, 1-28.
4. Cattell, R. B., *The Fight for our National Intelligence.* London: King, 1937, pp. 166.
5. Charles, E., *The Twilight of Parenthood.* London: Watts, 1934, pp. 226.
6. Duff, J. F., and Thomson, G. H., "Social and Geographical Distribution of Intelligence in Northumberland", *Brit. J. Psychol.*, 1923, *14*, 192-8.
7. Fisher, R. A., *The Genetical Theory of Natural Selection.* Oxford: University Press, 1930, pp. 272.
8. Ginsberg, M., *Studies in Sociology.* London: Methuen, 1932, pp. 211.
9. Ginsberg, M., *Sociology.* Home University Library. London: Butterworth, 1937, pp. 255.
10. Haldane, J. B. S., *Heredity and Politics.* London: Allen & Unwin, 1938, pp. 185.
11. Harrisson, Tom, "Notes on Class Consciousness and Class Unconsciousness", *Sociol. Rev.*, 1942, *34*, 147-64.
12. Himmelweit, H., and Whitfield, J. W., "Mean Intelligence Scores of a Random Sample of Occupations", *Brit. J. Industr. Med.*, 1944, *1*, 224-6
13. MacIver, R. M., *Society.* London: Macmillan, 1937, pp. 596.
14. Marshall, T. H., "Social Class", *Sociol. Rev.*, 1934, *26*, 55-76.
15. Pear, T. H., "Psychological Aspects of English Social Stratification", *Bull. John Rylands Library*, 1942, *26*.
16. Schumpeter, J., "Die Sozialen Klassen im ethnisch homogenen Milieu", *Archiv. für Sozialwiss.*, 1927, *57*, 1-67.

17. TAWNEY, R. H., *Equality*. London: Allen & Unwin, 1938, pp. 284.
18. THOULESS, R. H., *General and Social Psychology*. London: University Tutorial Press, 2nd edn., 1937, pp. 522.
19. WADDINGTON, C. H., *An Introduction to Modern Genetics*. London: Allen & Unwin, 1939, pp. 441.
20. WOOTTON, B. (*Ed.*), "Some Aspects of the Social Structure of England and Wales", *Adult Education*, December 1940, *13* pp. 97-116.
21. YOAKUM, C. S., AND YERKES, R. M., *Mental Tests in the American Army*. London: Sidgwick & Jackson, 1920, pp. 303.
22. WOOTTON, B., *End Social Inequality*. London: Kegan Paul, 1941, pp. 63.
23. PEAR, T. H., "The Social Status of the Psychologist and its Effect upon his Work", *Sociol. Rev.*, 1942, *34*, 68-81.

CHAPTER VIII

CONCLUSION

"I've seen a curious variety of people and ways of living altogether. Odd people they all are, great and small, very much alike at bottom and curiously different on their surfaces."

—*Tono-Bungay*, by H. G. WELLS.

In the previous chapters I have discussed at some length what it is we know about the factors which form the framework around which human behaviour as it develops in society is built up. I will now try to recapitulate some of the principal things which seem to emerge from this study.

Within recent years a good deal has come to light about the physical basis of inheritance. It has been shown that this resides in the genes, that the genes are responsible for the development of characteristics, that the genes are strung out along the chromosomes, and that the chromosomes, a different number of which is characteristic for different species, arrange themselves in pairs. Genes in the same position on the same pair of chromosomes are responsible for the development of the same characteristic, but the genes themselves are of two types—dominant and recessive. When they are dominant only one of the pair is necessary for the characteristic to develop, but when they are recessive both members of the pair need to be present before the characteristic appears.

On the basis of facts very similar to these Mendel developed his laws of inheritance, and the early geneticists attempted to apply these laws very widely so as to account for and predict the inheritance of a large number of physical characteristics. But soon it was found that the characteristics did not always appear in the offspring in exactly the proportions that were predicted. Three reasons have emerged to account for this fact. In the first place the genes themselves may sometimes change their character (mutate) in an unexpected and unpredictable way, with the result that they adopt a form different from that which, on theory, they ought to have, and thus influence the development of a particular characteristic in a different way. This mutation, which at first was completely inexplicable, is beginning to come under human control. It is at present possible to influence the mutation rate by subjecting chromosomes to X-ray, or other forms of short wave, bombardment. In this way genes have been turned from the dominant to a recessive form and also from a recessive back to the dominant. Furthermore it is found that the mutation rate seems to increase if the external environmental conditions are made very much less favourable to the organism.

At present, however, although it is possible to influence the rate of mutation it is not possible to control it nor to produce the mutation of any particular gene from one form which, let us say, is considered to be less desirable, to a form which is considered to be more desirable. Nevertheless, this line of experimentation opens up exciting possibilities. It might be possible some time in the future, when more is known about the precise conditions which are necessary in order to produce a desired change, so to influence the genetic make-up of an individual by artificial means as to compensate for an unfortunate combination of hereditary characteristics. Driven to its logical extremity the political implications of any such power are terrifying in their potentialities, but fortunately this is a problem which we do not have to face at present.

In the second place, dominance and recessiveness are not always found to be perfect. Sometimes a particular gene or pair of genes does not show its expected effect, because of the influence of other genes on the same chromosome, or sometimes on other chromosomes. The genes, therefore, can no longer be regarded as isolated factors: their full effect has to be considered in relation to other, maybe many other, genes. This is a serious limitation on the accuracy of mathematical predictions. Not very much is known about genotypic milieux at the moment, for investigators have not unnaturally preferred to work on those genes which give them the clearest and most predictable results. Nevertheless, before any wide application of genetic principles can be made, this is something about which much more needs to be discovered.

In the third place the characteristics which will finally emerge in an individual will depend not only on possible gene mutations and on the influence of genotypic milieux but also on the existence of particular environmental conditions. Through environmental conditions one is able so to affect the development of *axolotl* as to turn it into *amblystoma*. What heredity does in one case may be done by environmental agencies in the other: and these facts again open up startling possibilities in their application to human beings.

For many reasons, however, it is likely to be a long time (for which many people will be thankful) before there is any question of applying the results of research of this kind wholesale to human beings. The time may not be so long delayed before some kind of application can be made to physical characteristics, but on the mental side there are many more difficulties. These partly arise out of difficulties in diagnosis, and partly because of the quantitative and relative rather than qualitative and absolute aspect of most mental characteristics. Although it is true that there are certain quantitative aspects to most physical characteristics, although, for example, one can say that a person has a severe or a mild attack of smallpox, yet it is possible to establish a satisfactory dichotomy composed of those who have smallpox and those who have not. On the mental side, however, one cannot do this: even when one says that someone has no intelligence one means, of course, that he has less intelligence than the average, or (sometimes) less intelligence than oneself. It is true that one does not talk of a severe or a mild attack of intelligence, but one does try to express the amount of intelligence which a person possesses in the form of some quantitative scale by comparing him with other people. It may not be very accurately executed at the moment, but this is the intention. And so it is with other mental characteristics.

The problem, therefore, has to be approached from a rather different angle from that which is employed when most physical characteristics are under consideration: and the approach that is made has depended within recent years very largely on the device of the correlation coefficient, through which one may discover the degree of association that exists between different environmental circumstances, or different hereditary endowments, and more or less of the mental characteristic that is being measured. But before any great reliance can be put on the results one must be sure that the characteristics in question have been accurately measured. And when one investigates this problem one finds that, so far as most mental characteristics are concerned, the technique of measurement is in a very immature state of development, and that even in the case of intelligence, on which very considerable work has been done during the past forty years, the scores which are obtained from intelligence tests are not always as accurate as is needed for exact prediction. There are several reasons for this. If the test is one, as most of them have been in the past, which depends on words and on the use of words, it is clear that the facility with which a person can make use of words appropriately will depend, at part at any rate, on the training which he receives in the use of words, and this will turn on the literacy or relative illiteracy of his home and family environment. But even with those tests which are not linguistically biassed a person's ability to answer will still turn on the degree of his familiarity with the test situation, as well as on the level of his innate intelligence. Thus it is that town children and country children have each been shown to be more intelligent than the other, according to the type of test that has been used to measure their intelligence, and the typically rural or urban situations on which the tests have been founded. And even those tests which are neither situational nor verbal but which are based on the abstraction of formal relationships cannot be shown to be entirely culture free, for again the technique of abstracting relationships is something which has to be taught, or which may be picked up more readily and easily in one environment than in another.

With these limitations in the accuracy of the measuring instrument, one must be careful not to read too much into results which differ by only a few points of intelligence quotient. But one of the most hopeful lines of investigation would appear to lie in the study of twins. When groups of fraternal twins are compared with groups of siblings the results of intelligence tests are strikingly similar. This is what might be expected, since the fraternal twins are no more likely to have a similar heredity than ordinary brothers and sisters. When groups of identical twins are compared with groups of fraternal twins or of siblings the results are much closer in the identical twins than they are in either of the other two groups. But even here in one typical investigation the average difference between pairs of identical twins was found to be as much as 5.9 points in I.Q., and since identical twins have the same heredity this difference must either be due to inaccuracies in the measuring instrument, or to differences in the environments to which the different members of the twin pairs were subjected in their homes, or both.

A certain number of identical twins have been investigated who have been brought up not in the same homes but in widely different environments, and these are of special importance for purposes of comparison with those in whom both heredity and environment are similar, or with those fraternal twins whose heredity is different. And in these cases it is found that the identical twins who have been separated since infancy are as different as fraternal twins who have been brought up together. In some of these cases, too, the difference between the different pairs of identical twins is sufficiently great for one to be able to eliminate the inaccuracy of the measuring rod as the sole explanation of the difference. Similarly, the difference between the scores of some of the different pairs of fraternal twins is sufficiently great for one to be able to eliminate the inaccuracy of the measuring rod as the sole explanation of their difference. Thus, so far as the intelligence of twins is concerned, one may conclude (as one concluded from the experiments on physical differences) that differences which may be produced by hereditary differences in one case may be equally well produced by environmental differences in another.

The same degree of physical or of psychological differences may be produced either by hereditary variation or by environmental alteration. But can this be true of sex differences? Have we not in sex a fundamental and qualitative hereditary difference between people, which no amount of environmental manipulation can overcome? Here again, however, an inspection of the evidence must lead us to the conclusion that even sex is a quantitative rather than a qualitative characteristic.

The primary hereditary difference between one sex and the other lies in the fact that one pair of chromosomes has a different form in one sex from that which it has in the other. In mammals the female has two "X" chromosomes while the male has an "X" and a "Y", the "Y" chromosome playing no significant part in the determination of sex. Now from one generation to the next the "X" chromosomes may pass from one sex to the other and determine a different type of sexual development. Thus a male receives his "X" chromosome from his mother. In his mother it was one of a pair which helped to determine her sex as a female: in him it is the only one which determines his sex as a male. Similarly a daughter receives one of her "X" chromosomes from her father. In him it determined his sex as a male: in her it is one of the pair which determines her sex as a female. The function of the "X" chromosome, therefore, varies according to whether it is alone or possesses a partner.

But sex does not depend on "X" chromosomes alone: it is rather the resultant of a balance between "X" chromosomes and autosomes. The autosomes are male determining, the "X' chromosomes female determining. Where there is only one "X' chromosome the male-determiningness of the autosomes is stronger than the female-determiningness of the "X" chromosome: but where there is a pair of "X" chromosomes the position is reversed. Furthermore by X-ray (or other short wave) bombardment part of "X" chromosomes or parts of autosomes can be destroyed, and when this is done it is found that the greater the amount of "X" chromosome destroyed the more masculine the developing individual becomes, and correspondingly the greater the amount of autosome that is destroyed the more feminine the developing individual becomes. All degrees can be, and have been, produced between super-femininity through normal femininity, through intersexuality, through normal masculinity to super-masculinity. Thus it is, therefore, that sex must be regarded as quantitative rather than as qualitative.

In mammals, however, sexuality depends not only on the chromosomal balance between autosomes and "X" chromosomes, but also on the secretion from particular endocrine glands. Normally, after a period of development in an "indifferent" condition those individuals who possess only one "X" chromosome in their germ cells begin to produce testicular glands and then to be affected by the secretions from these glands. At a slightly later period of development in an "indifferent" condition those individuals who possess two "X" chromosomes in their germ cells begin to produce ovaries and then to be affected by the secretions from them. Now just as the development of different sexual characteristics may be affected by altering the balance between "X" chromosomes and autosomes, so also the development of different sexual characteristics may be affected by transplanting testes from the male into a female body from which the ovaries have been removed, or *vice versa*. In mammals, therefore, we have not only the chromosomal balance to consider but also the amount of secretion from testes or ovaries, and the masculinity or femininity of the developing individual will depend on the kind and extent of the flow of secretion. Once more the quantitative rather than qualitative character of sex becomes clearly apparent. Finally, the development of particular physiological sexual characteristics has been shown to depend in some organisms on the kind of external environment in which they are reared. Thus in one external environment they will develop as females, but in another they will develop as males.

With so much possible modification in the appearance of physiological sex differences one would expect to find the influence of the environment to be still more strongly marked when one turns to the consideration of mental differences. And this is found to be so. Men, it is true, have achieved greater eminence in Western civilisation than have women, but there is nothing to indicate either that they possess greater native intelligence nor that the range of their intelligence is any wider than it is in women. Furthermore, the anthropological evidence indicates that the rôles assigned to men and women have been widely different in different societies, and that there are even some in which the social expectation of those rôles is precisely the opposite of what it is in our own society. To account for the psychological differences in the attitudes of men and women in our own society, therefore, it is probably more profitable to consider the kind of social expectation which exists. It may be possible, as Adler tries to do, to explain the different types of behaviour in terms of the relative strength of feelings of inferiority, but it is unlikely that this is a complete explanation. At all events there is little doubt that women behave as women and men as men not because of a fundamental and qualitative difference in their physiological or psychological make-up but because they are forced into these different ways of behaving by the pressure of social expectation, custom and training. It is true that a particular chromosomal balance or a particular kind of glandular secretion may make a man or a woman more or less masculine or feminine than the average, and that this may affect favourably or unfavourably their behavioural reactions to the cultural pattern of the society into which they are born, but the point is that taking groups of men or of women as a whole the masculinity or femininity of their behaviour is likely to be more greatly affected by the expectations of the society into which they are born than by physical determinants.

On turning to the question of racial differences we are now prepared not to overestimate the value of hereditarily determined characteristics. But in addition to this the picture presents various new and interesting features. Races have been defined in terms of their similarity or difference in the possession of hereditarily determined characteristics, but when we consider the possible criteria in rather greater detail we find that practically all of them are variables rather than absolutes. This in itself is not an insurmountable difficulty, but it becomes so when we find that when we take each of the most obvious physiological determinants in turn our racial classification alters according to the criterion we happen to choose. Thus on the racial problem we have not even a clear-cut starting point. Then again environmental influences can be shown to affect many of the criteria we decide to use. This is even true of some of the physical criteria, including cephalic index, but its effect is much more marked if we attempt to include among the criteria for distinguishing between different races intellectual characteristics as well. If we do this we are brought right up against the problem of the effect of environment on intelligence test scores, and we then find that, apart from the effects we have already considered, the scores vary according to the educational efficiency of the area in which the person was educated, according to the length of residence in a more favourable area, and according to the particular customs in different areas which affect the motivation of the person doing the test.

When we consider a person's racial background, therefore, and the effect it is likely to have on his subsequent behaviour, we are forced to turn much more to the effect of different environmental conditions. And when we come to the question of national differences and of class differences we have to leave behind purely genetic considerations. Here we find that none of the purely objective criteria, such as language, religion or territory, is by itself a sufficient explanation of the development of nationality, and hence of national character. Consequently we have to turn to a complicated mixture of factors which depend on the possibilities which lie open to a group of people and which includes their geographical environment, their type of population, their cultural pattern, their political organisation, and so on, and to examine these from a historical point of view before we are in a position to examine with any thoroughness exactly what is meant by national differences. None of these factors depends primarily on genetics, yet taken together they form the principal moulding force in a person's behavioural development.

They form the principal moulding force in all communities, though in some communities other forces act on him as well. Such, for example, are those which may arise through differences in social class. In our own society certain advantages appear to arise if one possesses a particular combination of wealth, birth, education, occupation and leisure habits, and certain disadvantages with a different combination.

These, then, are the factors which form the framework of human behaviour. Purely hereditary endowment, whether of physical, mental, sexual or racial characteristics, it is even now possible to alter by environmental manipulation, though it is not yet possible adequately to control them. And as for the purely environmental factors of nationality and social class these should be at least as easy to control. At present people start their development along particular lines because of the hereditary and environmental framework with which they are endowed. In the future, perhaps, they may be enabled to start their development along more favourable lines through the human manipulation of these framework factors.

INDEX OF AUTHORS

INDEX OF SUBJECTS